THE F*CK IT DIET

CAROLINE DOONER

EATING SHOULD BE EASY

This book contains advice and information relating to health care. It should be used to supplement rather than replace the advice of your doctor or another trained health professional. If you know or suspect you have a health problem, it is recommended that you seek your physician's advice before embarking on any medical program or treatment. All efforts have been made to assure the accuracy of the information contained in this book as of the date of publication. This publisher and the author disclaim liability for any medical outcomes that may occur as a result of applying the methods suggested in this book.

HarperCollins books may be purchased for educational, business, or sales promotional use. For information, please email the Special Markets Department at SPsales@harpercollins.com.

FIRST EDITION

DESIGNED BY WILLIAM RUOTO

Library of Congress Cataloging-in-Publication Data

Names: Dooner, Caroline, author.
Title: The f*ck it diet: eating should be easy / Caroline Dooner.
Description: First edition. | New York, NY: Harper Wave, [2019]
Identifiers: LCCN 2018041370 | ISBN 9780062883612 (hardback)
Subjects: LCSH: Weight loss—Psychological aspects. | Diet—Psychological aspects. | Diet therapy. | Nutrition. | Women—Health and hygiene. | BISAC: HEALTH & FITNESS / Diets. | SELF-HELP / Eating Disorders. | SOCIAL SCIENCE / Feminism & Feminist Theory.
Classification: LCC RM222.2 .D674 2019 | DDC 613.2/5—dc23 LC record available at https://lccn.loc.gov/2018041370

19 20 21 22 23 LSC 10 9 8 7 6 5 4 3 2 1

I DEDICATE THIS BOOK TO CHEESE—
I WILL NEVER FORSAKE YOU AGAIN.

CONTENTS

ONE

WHY ARE WE SO ADDICTED TO FOOD?

TWO

HOW THE HELL DO I ACTUALLY DO THIS?

This book is meant for chronic dieters. I am not a medical professional, and if you are engaging in any extreme restrictive eating or self-harm, you must seek professional medical help. This book cannot stand in for eating disorder treatment and is not meant to treat any physical or mental conditions.

I will share the work of many doctors, nutritionists, and weight and metabolism scientists to help assure you that I am not a lone, crazy brownie eater, out to ruin your health and happiness . . . but, again, I'm not a doctor. This book isn't medical advice. Okay. You get it. Moving on.

THE
F*CK IT
DIET

INTRODUCTION: THIS IS NOT A DIET BOOK

If you've read as many diet books as I have, maybe you've noticed how similar the introductions are. It's normally a sensationalized promise that *this is the diet you've been waiting for.* And it goes a little something like this:

INTRODUCTION TO THE DIET YOU'VE BEEN WAITING FOR

You've been going from diet to diet, and nothing has worked, you're still ~~worthless and ugly~~ fat and unhealthy. But now with this groundbreaking ***and*** *ancient way of eating, you finally have a proven, scientific, and simple plan to unlock all of your dreams of being ~~beautiful and loved~~ fit and healthy.*

The best part is this: if you follow the plan exactly the way we have laid it out in this book, you will never crave food ever again! ***Ever! No more food cravings.***

Take it from me, craving food should not have to be a part of your life.

All of those other diets you've tried in the past didn't work because you weren't eating the ***right*** *kinds of foods in the* ***right*** *amounts.* ***Of course*** *they didn't work.*

Well, with our program, your body will begin to work ***so well*** *that you'll never think about food again.*

This isn't a diet, it's a ***way of life****. It's a* ***lifestyle change****. I know that those* ***other*** *diets you went on said the same thing, but they were lying. Those were* ***totally*** *diets. This isn't.* ***This diet is totally NOT a diet,*** *it is a way of life.*

Are you ready? Let's turn the page to the first chapter, where I'll tell you why everything you are currently eating is bad for you.

You read this and think, *THIS IS IT! Finally, a way to get rid of my cravings. I'm so sick of being hungry.* You get rid of everything in your kitchen, you stock up on the allowed foods, and you start devotedly eating exactly what they prescribe, getting a little adrenaline high every day that you successfully stick to the rules. You're so excited to have finally found this scientific *and* ancient diet that will take away your weak, human desire for food. You have been so frustrated with your hungry humanity that you are willing to do *anything* to be less of a burden to society with the space you take up.

You devote three *extremely* committed months to this scientifically proven way of eliminating your body fat and your cravings and . . . *voilà!* It goes so well! You are toned, energetic, and as happy as ever, eating your perfectly portioned, perfect food. You start forgetting to eat and consider becoming a robot who only needs to take pellets occasionally.

The best part is that all your relationships flourish because, now that you don't need food, you are *so* personable. Everyone loves you even more than they did before, and they all want you in their lives. You go out to lunch with friends and just smile at them while they eat and you think about how great your life is. Everyone thinks you are funny and beautiful and perfect and they wish they could be more like you.

You also become rich and you're never, ever bored.

Don't you worry. This book will be nothing like that. Because The Fuck It Diet is not really a diet.

I used to be an extremely devoted dieter (when I wasn't bingeing), and those diet book introductions were always sooo exciting to me. *I will DO this. I will do this right. And I will finally make my life AMAZING.*

And I *would* do it. Until I eventually failed and started the binge/repent yo-yo, or until I went on a bender, or replaced it all with another, better diet.

My dieting started at fourteen, when I realized my shorts were really tight and my face was becoming oilier and puffier by the day, and I had to go to Nordstrom's to buy bras in a size E in the brand that Oprah recommended because Victoria's Secret bras were too small.

*I **have** to fix this. . . . I guess my days of eating are behind me.* So for the next ten years I was either "on a diet"—obsessed with following the rules perfectly—or "off a diet," because I was bingeing and feeling out of control and horrible about myself.

I tried the Atkins diet, the South Beach diet, the insulin resistance diet, the pH diet, the blood type diet, the Rosedale diet, the raw vegan diet, many versions of the Pray to God to Be Skinny diet, The Secret™ (not a diet but you can make anything into a diet, especially new age self-help), the "I'm Going to Listen to My Body SOOOO WELLLL" diet (also known as the obsessive version of intuitive or mindful eating), the *French Women Don't Get Fat* diet (which is a hybrid of the intuitive eating diet and the coffee and wine diet), the paleo diet, the GAPS diet, anddddddddd . . .

Boom. Epiphany. It hit me, on my twenty-fourth birthday, after I ate nine squash "pancakes" and twelve sugarless almond flour "cupcakes" that I made for myself and that nobody else would eat. I had a legit spiritual epiphany, with my stomach distended and my heart palpitating. I stared at myself in the mirror of my crumbly little bathroom on the Upper West Side of Manhattan, and as if I was in some sort of not-that-funny romantic comedy, I spoke to myself out loud. "What are you *doing*? *Are you going to do this for the rest of your life?*"

I'd spent the past ten years truly hating my body, constantly disgusted with myself, and wanting to be skinny more than anything else. I spent years obsessing over diet rules, planning what and when I could eat next, and counting up calories and carbs. I spent all of my energy trying to control my weight and salvage my health, but still, no matter how hard I tried, or how important dieting was to me, I was bingeing. I felt completely out of control for years on end.

I was petrified of carbs and sugar, and being full, and absolutely everything I did was for the purpose of trying to weigh less. Every day was good or bad based on the number on my scale and what I had eaten. And I truly believed I was doing it all in the name of health, because in my understanding, health and weight were synonymous.

Also, every fantasy I had was basically just me being skinny and pretty and *maybe dating Prince Harry, I don't know.* But definitely skinny and pretty, as if *those* were my actual dreams and as if *that* was my actual purpose. As if being skinny and pretty was the thing that would make me happy.

As for my *actual* distant buried dreams? *Well, if I can be skinnier, then they'll* ***finally*** *work out. Once I am skinny for good, I can finally take myself seriously.*

But even the times when dieting "worked" and I was actually skinny, it was never, ever enough. I didn't *feel* skinny, or worthy, or confident. And the moments that I *did* feel skinny? I was mostly panicked that the skinniness wouldn't last, and that made me even more obsessed with dieting.

I'd spent ten years thinking that skinniness would make me like myself. I thought that skinniness would make me happy. And that is a method for happiness that *doesn't last.*

Skinny doesn't create happiness—just ask any model or any

"successful" dieter. Sure, once you reach your goal weight, you think you're happy for a moment. If you're reading this book, you already know that it doesn't last. Changing this search for happiness-through-skinniness-and-beauty to something more real, attainable, and life-affirming is a big part of this book.

But the part we are going to focus on first is the fact that *diets don't even work*. The way we try to exert control over our bodies is biologically flawed and set up to fail from the start. When we try to override our survival response, our survival response *wins. Every time.*

We will get into it all. But first let me tell you what I did after my bathroom mirror epiphany. I decided to learn to eat normally—and I finally understood what that meant: I had to eat a lot more than I ever thought was okay before. I decided to give in to all of the foods I was afraid of, and *all* the hunger I had been trying to repress for the past ten years. And I mean all of it. *There was a lot of hunger.*

I also decided to research all of the reasons dieting doesn't work. I armed myself with every bit of scientific information I needed to keep myself trusting that *not* dieting was the right path. I found a whole movement dedicated to educating the world on why the way we approach health and weight loss is deeply flawed. I learned about all the ways I was actually screwing myself over with diets—on a biological, chemical, and hormonal level.

But most importantly, I decided I was going to learn to like and accept myself at whatever weight I ended up. I really wasn't sure what that weight was going to be because I'd been yo-yoing up and down *many* times every year for the past ten years. I thought my weight might end up at the top of my yo-yos, my highest weight, where I'd always felt like a serious failure. There was *nothing* more panic-inducing to me than seeing my weight up there, but I was choosing to change my priorities, big-time. I decided, *Fuck it.* Seriously, *fuck it.* I was too miserable not to do this.

I was going to learn to like myself wherever my weight fell, because I could not spend one more day fighting myself, waiting for that magical, elusive day when I would finally become permanently skinny and content. I knew that this was the only way out of the trap I was in—and *The F*ck It Diet* was born.

WHO THIS BOOK IS FOR

This book is for chronic dieters. This book is for people who are ready to learn why their diets haven't worked, and *why* what we've been taught about food and health hasn't worked. It's for the people who have been on every diet, who have spent hours worrying about and micromanaging the minutiae of the calories or toxins in the food they are eating—and don't want to do it anymore.

It's for people who have spent years seeing their worth through the lens of what they ate that day and what they weighed that morning, who have gone from diet to diet hoping that the next one would be the answer. It's for people who didn't even realize how miserable they were, because they were too busy praying that *just maybe* this time they'd lose enough weight to like themselves, and then all this misery would be worth it.

If you feel perfectly fine with the way you eat, exercise, and relate to your body and weight, you probably don't need this book. But for those of you who are sick of being stuck in an abusive relationship with diets, who want a different relationship with food and with your body, this book is here to tell you that there is a way out.

I am now completely easy-breezy with food, which I genuinely didn't think was possible before. Since going on The Fuck It Diet, I barely even think about food anymore if I'm not hungry, which was also something I thought was a myth. For so long, I believed that the cure for binge eating and food obsession was *better willpower.* I thought that if I could just diet successfully, like I did those first few times, and then keep doing it forever, I would finally be healed, happy, and—most importantly—thin and beautiful.

The irony is that restriction and dieting cause a very real food addiction that cannot be cured with more dieting and more restriction. We are physiologically and psychologically wired to be food addicts when our bodies sense there isn't ample food. It's chemical and hormonal and completely inescapable.

No matter what you weigh, dieting is ruining your metabolism and your ability to listen to your body. We are going to talk a lot more about weight science, and the reasons your health and weight aren't as connected as you've been taught, and how diet culture has basically put you at war with yourself.

This book can benefit any person, of any gender, at any weight, who struggles with food and body image. But because I am a woman who had to figure out why I was so scared of being too big in this world I live in, this book is inherently a feminist response to diet culture. The insidious societal causes of our dysfunction with food and weight can't be ignored. So, to the women who believe they must be tiny and toned in order to matter and be respected, I say "*Fuck that.*" You are allowed to eat the whole sandwich, and you are allowed to take up as much space as your body needs.

I have worked with well over a thousand women (and some men) through The Fuck It Diet in my group programs, as well as one-on-one. And over and over again—to everyone's surprise—I've seen that allowing all foods is the *only* thing that heals food obsession and bingeing. The most common fear is that once they start eating, they will never stop. And every single time, people are in awe of how their appetites completely change once they are actually fed—once you give yourself permission to eat, bingeing disappears. No superhuman willpower required.

The reason The Fuck It Diet works when no diet, self-help guru, or mindful eating does is because it tackles two things at once: the biological reasons that keep people obsessed and bingeing, and at the exact same time, the mental, emotional, and *cultural* reasons that we become obsessed with food and weight in the first place.

In this book I'm going to share my experience, their experiences, and the science that *backs up* our experiences and explains why not-dieting actually works. These lessons were hard to learn, but once I learned them, they became so obvious and logical. Now I wonder how I ever believed that restriction was the answer.

ONE

WHY ARE WE SO ADDICTED TO FOOD?

LET'S GO ON A FAMINE

Do me a favor and imagine that you are in a real-life famine and you have access to very little food. Just imagine what would happen.

Immediately, everything in your life would become about food. Everything in your body would be telling you both to ration what you have *and* to eat a lot the first chance you find enough food. You would be constantly looking for more food. You'd maybe start searching for crops that weren't destroyed. You'd hunt rabbits. You'd forage. And you would quickly become very resourceful with the food you did find.

There would be a surge of adrenaline when this restriction and food search begins—it's slightly euphoric—giving you enough energy and hope to scavenge for food. But at the same time, your metabolism would slow down so it can resourcefully use and store the nutrients you *are* eating. As you are forced to eat less, you would probably lose weight, but at the same time your metabolism would slow down so you don't lose *too* much *too* fast—because if you used too much fuel too fast, you'd die.

After you've been hungry and rationing for a while, always eating what you can when you can, you may finally come across more substantial food. Maybe you spear a boar. Maybe you steal some loaves of Wonder Bread from a rich family in the village. Whatever. The point is: you find more than a handful of food, and everything inside you overrides whatever rationing willpower you've had so far. You *eat it all.* You eat as much as you can get. You *feast.* And if you tried to stop yourself halfway through, you probably wouldn't be able to.

That's what your body is wired to do for survival. It's a *good thing.* Your body's only job in a crisis is to help you store nutrients and fuel in your body for the days and weeks to come. It gives you some energy, though you'll still be operating at a lower metabolic rate than normal if the feasting isn't able to continue. You're still in a famine, even if you just ate two loaves of Wonder Bread. Your body knows you're still searching for food, constantly.

To stay alive, you will have to *keep* eating as much as you can when you find it, and your metabolism will remain low while you do, ensuring that you stay alive.

There are two possible endings to this famine:

FATE #1: THE FAMINE NEVER ENDS. As you use up all your food stores, you stop being hungry at all, because your body believes there really is no food, so it is not going to keep using precious energy to send hunger signals. You live for a little while like this, in deteriorating health, then you die. *And* you can and will die of starvation *even if you still are not emaciated*, because starvation weakens your muscles and heart regardless of your weight.[1]

FATE #2: YOU FIND ENOUGH FOOD TO KEEP YOU ALIVE BEFORE THE FAMINE ENDS. But before it fully ends, every time you find food, you feast. As you should. Your body stores those calories as fat to help you rebuild and repair your body, and to protect you in case you find yourself in a famine again.

In between these necessary and helpful feasts, you are hungry and still fixated on finding and eating as much food as you can, when you can. Of course.

Before the famine is over, other things happen as you go through feast-and-famine eating: your hormones stop working properly and your sex drive drops (no use having children in the middle of a famine!), you're irritable, and that adrenaline high is wearing off. Your body is trying to conserve energy, so your metabolism is low and your energy may come mostly from spikes of adrenaline and stress hormones.

Maybe thanks to some sort of manna, or because you found a more bountiful terrain with fish and mangoes and brownies, you live and the famine eventually ends.

Once there is food, you are going to eat as much as you can, for a long time. You will gain weight, and it will be **awesome**. Your body will take some time regaining strength and vitality. You will be tired for a good chunk of time, while your body slowly repairs the parts of you that were sacrificed in order to keep you going during the famine.

During your recovery from the famine, every time you see food, you're gonna eat it. Because *of course you are*. There was just a famine! You were starving for a half a year! Or five years! Your body is not convinced that there isn't another famine right around the corner, so you're going to be eating a lot for a while. You're going to need to rest for a while. And you *will* gain weight during this recovery, as you should.

Once your body is fed for a long time, and not worried about any more famine, you will slowly come back to normal. Food won't be as stressful. You will slowly trust that there is enough food again, and your body's metabolism will eventually normalize. Your appetite and desire for food will eventually normalize, and your weight will eventually stabilize—maybe slightly higher than it was before, just because of a fear of future famines, or maybe not.

I'm sure that you've made the connection by this point, but let me spell it out anyway: dieting is putting your body through a famine. That may sound like a stretch, but it's not. Not at all. You'll say, "No no, I eat plenty, even when I'm on a diet." Or you'll say, "Um, I am bingeing all the time, there is no way my body doesn't have enough food."

It doesn't matter. If you are still eating, but just not *quite* eating to satiation, or if you've been yo-yoing between dieting and bingeing, the body reads that as a famine state. Let me say that again: **If you are yo-yoing between dieting and bingeing, you are putting your body through a constant crisis.**

This is a crisis and survival state. Before our current diet culture—which, by the way, is only decades old—the *only* reason you ever would have eaten less than sufficient food would have been if there was a shortage: a famine. Eating less than you are hungry for triggers your body's survival mode, changing your hormones and brain chemistry, which then lowers your metabolism and makes you biologically *obsessed* with food. The mental fixation is actually *caused* by the physical restriction.

Food fixation and bingeing are both caused by your body trying to force you off your diet/famine for your survival. If you trusted the food your body was forcing you to eat, followed your natural hunger, and let yourself recover, you'd recover relatively quickly. Your body knows what to do. It might take a few weeks or months, but then your appetite, metabolism, and weight would eventually stabilize.

But we never let ourselves do that. We *don't* let ourselves eat a lot because we don't trust our appetites or our weight. We have been told that eating a lot is bad, and a sign that we are *surely* food addicts. In fact, we *fight* our natural urges to eat a lot and to rest, fearing that we are lazy and irresponsible. We trap ourselves in this famine state, and so the food fixation continues. Then we become one of those old ladies in the nursing home worried that their pudding is going to make them fat.

When you restrict, your body is wired to compensate for the lack of food, slow down your metabolism, fixate on food, and hold on to weight. When your metabolism is compromised, your body is going to, basically, slowly deteriorate your health in order to keep you alive for as long as it can, in the *hopes* that one day you will be able to eat a lot again and give your body a chance to repair and recover.

If you are obsessed with food, you have triggered a famine state. If you are bingeing, you are in a famine state. This is true no matter how much you weigh, or how much you are *sure* you are already overeating.

You can put your body in crisis mode *even if* you are only restricting "a little." If you are keeping yourself hungry often, it'll happen. It's also very important to note that your body can be in this state *even if you are not very skinny.* Many people who don't *look* underfed are in a famine state. This biological and metabolic phenomenon will happen whether you are tiny or fat. The body will need more fat while it recovers no matter what, as a sort of insurance policy.

It's hard for us to believe that the cure for our food addiction could possibly be through eating *more* and letting our body heal from the reactive and food-obsessed famine cycle. We are too afraid of food and calories and weight, so we never recover, and our obsession and bingeing continues. The yo-yo gets worse, our metabolism stays suppressed, our brains fixate on food—and our body puts on weight at any chance it gets.

We are convinced that our main issue is food addiction and overeating but we are completely oblivious to the fact that it all stems from restriction. In fact, we can argue that fat bodies are wired to resist diet/famine even *better*. Your body doesn't want you to lose weight, for fear of an upcoming famine. And through this lens, a fatter body is better wired for survival.

The body does not like it when you try to control food intake. It doesn't understand you are trying to fit into absurdly small jeans. It fights back against famine and restriction for your survival, and the more you diet, the harder it fights back.

THE MINNESOTA STARVATION EXPERIMENT

During World War II, there was a starvation study conducted by Ancel Keys at the University of Minnesota. He wanted to learn how best to rehabilitate starving people after the war—so first, he had to starve people.

Over four hundred conscientious objectors applied to participate in the study as an alternative to fighting. Only thirty-six men were chosen: those who were the most physically and mentally sound, and who were the *most* willing and aligned with the goals of the experiment.

The men stayed together in dorm-like rooms connected to the temporary laboratory. They were allowed to leave, but the compound was their home base. For the first three months, the men ate normally while their health was closely monitored. They were fed around 3,200 calories a day, which was considered a normal amount. (Because it is.) They took jobs on the compound and walked around twenty-two miles a week.

Then, for six months, their calories were dramatically cut—in half. They were only served two meals a day, which worked out to roughly 1,600 calories total. The participants were encouraged to keep up their walking.

In this experiment, 1,600 calories was considered "semi-starvation," which is really horrifying when you realize that this is the same "conservative protocol" used by the FDA to "combat obesity." You've probably seen that calorie number floating around fitness magazines and doctor-prescribed diets. These days, 1,200–1,600 calories is considered an acceptable daily amount of calories for men and women.

Men often run on more calories than women because of both size and muscle composition, but 1,600 is too low for anyone. In fact, even the new 2,000-calorie recommended daily intake "is only enough to sustain children,"[2] according to Marion Nestle, PhD and professor of nutrition and food studies at NYU. Let that sink in.

So on only 1,600 calories, the participants' strength and energy immediately began to decline, and they said they were constantly tired. Then apathy set in. They had all been strongly opinionated conscientious objectors, but now they didn't really care about any of the things they used to care about. Next, sex and romance lost its appeal.

All their thoughts became about food. They became completely fixated on thinking, talking, and reading about food. (Sound familiar?) Some began to read and stare at cookbooks for hours, mealtimes became their favorite part of the day, they'd be irritable if they weren't fed on time, and even though their food was bland bread, milk, beans, or vegetables, *they* thought it tasted amazing. Many men would mix their food with water to prolong the meal, or drag out meals for two hours, or sneak food to their rooms to savor it slowly.

The men had access to unlimited coffee, water, and chewing gum in between meals, and the men became addicted—some of them chewing forty packs of gum a day, and having around fifteen cups of coffee.

The men, who had been, on average, healthy and muscular to start, became extremely skeletal during those six months. Their heart rates slowed way down, and the men were cold all the time—both symptoms of low metabolism and the body trying to conserve energy. Their blood volume shrank, their hearts shrank, and they developed edema and retained water. Their skin became coarse, they were dizzy, lacked coordination, and experienced muscle soreness.

On the bright side, the whites of their eyes became brilliantly white because all their blood vessels shrank! So if you want to have beautiful porcelain doll eyes, starve yourself. You'll just have to deal with lots of other horrible problems.

Next, they started sneaking food from off-site. Remember, these men had been chosen *specifically* because they were the most willing and likely to comply with the experiment. But they *still* started cheating with extra food off the compound. In fact, the cheating became such a huge issue that the men were required to have chaperones every time they left. Three men completely pulled out of the experiment.

These men were also profoundly psychologically changed by their restrictive diets. A few weeks into the experiment, one man started having disturbing dreams of cannibalism. Then he cheated on the experiment by going into town and devouring milkshakes and sundaes. When the head of the experiment confronted him, he broke down crying and threatened his life. He was discharged and sent to a psychiatric hospital, where after a few weeks of being fed normally, his psychological health went completely back to normal (!!!). LET THIS SINK IN! All this man needed to regain his sanity was more food.

Yes, this man was an extreme case, but *all* of the men became anxious and depressed. One man recalls snapping at his good friend in the experiment nearly every day, and having to apologize often for his irrational outbursts.

And the weirdest part of all: even though these men had become extremely emaciated, they did not perceive themselves as being excessively skinny. Instead, they thought *other people were too fat*. They were experiencing body dysmorphia, which is a phenomenon experienced by people with eating disorders where people see their bodies as a different size or shape than they actually are. It's assumed that eating disorders could be a result of body dysmorphia, but these men didn't even want to lose weight in the first place. They were experiencing psychological body dysmorphia *just* from the physiological effects of starvation. I can't explain that one to you. But it's eye-opening.

So what do you think this means for a culture obsessed with controlling the food we eat and the way our bodies look? It does not bode well. Dieting and restriction messes with our brain chemistry big-time. It fucks with our mental health and takes over our minds until food and weight are all we can think about. We deserve better because *this isn't working*.

REHABILITATION

The purpose of this experiment was to see how to *rehabilitate* people who were starving, and the goal was to figure out how best to help them recover. These dramatic physical and psychological effects weren't even what the study was meant to focus on. The semi-starvation phase of the study was actually just to get the participants to the place that they needed to rehabilitate from.

When Keys started re-feeding the participants, he only increased their food a little bit, assuming at first that slowly re-feeding would be the healthiest method—some by 400 calories, some 800, some 1,600. The group whose food was increased by 400 and 800 calories had no improvement at all. He gave them supplements and protein shakes. They still didn't improve. The *only thing that worked was more food.* And lots of it. Upping their calories *above* what they ate before the experiment had an immediate positive effect.

However, for many of the participants, the emotional disturbances of starvation lasted throughout the rehabilitation process, and some reported being even *more* depressed and anxious during the re-feeding and rehabilitation than they had been during the restriction. That's important information for us, because it means that—hormonally and chemically—it can be a very bumpy road while you re-feed yourself after famine and dieting.

Only twelve men stayed for some extra months after the end of the experiment for what Keys called "unrestricted rehabilitation." On average, these men ate 5,000 calories a day, but sometimes as many as 11,500 calories a day. They often talked about a hunger sensation they couldn't satisfy, no matter how much they ate or even how full they were.

The men said there were lingering effects of this experiment, and many of them had recurring fears that food would be taken away from them again. Three of the men became chefs—all men who had no real interest in food or cooking before the experiment.

Many of them said they were very hungry and fixated on food for months *or years* after the starvation experiment. And in my research of this study, I've read mentions of the therapeutic effects of many, many milkshakes. That's the 1940s for you.

WHAT DOES THIS MEAN FOR DIETERS?

I mean . . . you see the problem, right? You see that mainstream recommended weight loss and "weight maintenance" diets—

which recommend anywhere between 1,200 and 2,000 calories a day—are right around how many calories these men were eating to induce biological starvation responses and deep, lasting fixations on food? You see how extreme the physical and mental consequences were on a diet of 1,600 calories a day? How everything in these men's bodies and minds screamed for food, and how in the end, the only cure was lots and lots of food, for a long, long time?

What these men experienced is nearly identical to what people experience on diets, and what they experience when trying to get their body out of diet crisis state. When you diet, even if it's just a little bit, even if it's a seemingly reasonable sixty-day plan you found in *Shape* magazine, you put your body into a reactive, food-obsessed survival state. Your fixation on food is not happening because you are lazy or irresponsible—it's an inescapable protective measure meant to keep you alive.

And for those of us who have a lot of trouble staying on a diet, even for just a day? Congratulations: that's actually a *good* thing! "Successful" calorie restriction has immediate and dramatic physical and mental effects. If those men hadn't been so closely monitored and controlled, they would have gone off their "diets."

Staying on a diet is at odds with our biology. But the saddest part of our diet-centric culture is that when our bodies force us off our diets, we keep forcing them back on. To become normal with food, you have to *deliberately* step out of this cycle, and get your body out of this crisis and survival state and back to some sort of normalcy.

WHAT IS NORMAL EATING?

Back before The Fuck It Diet, I was so far from normal eating and *so* fixated on food and weight that I didn't even know what it was supposed to look like. I would look at people who didn't overthink food and think, *Well, I guess they are just lucky to not have a food addiction.* I didn't realize that my "food addiction" was biologically driven, and that I was constantly making it worse by every diet I went on.

I didn't realize that, in a way, we are *meant* to be fixated on food. Because food is a fundamentally important part of staying alive, so when the body senses that food access is scarce, our food fixation increases. Thankfully the reverse is also true. Once the body knows it will be fed, it can calm down. Hallelujah.

Here are some things you'll experience when you are not stuck in the food survival state anymore, and become a "normal eater":

» You can go through your day and pretty much only think about food when you are hungry.

» You will have a strong healthy appetite for lots of food, yet your weight will stay stable because your metabolism isn't compromised from dieting.

» You eat what you crave, but you crave what you need. Sometimes salads, sometimes a cookie, sometimes fruit, sometimes steak, etc.

» You can eat a meal and stop in the ballpark of satiation and fullness without overthinking it.

» You can eat when you're distracted or tired or stressed or sad and *still* stop once you get full.

- » You will have a strong sense of what food you want, when, and how much, but it won't be *that* important that you follow your cravings perfectly, because *life is too short to obsess over food.*

This list is just a taste of what can naturally happen when you finally get out of the biological famine state. Ironically, it takes a good amount of relearning before eating becomes easy. But you *can* do this. And by George—whoever George is—I will help you get there if it is the last thing I do.

THE BIG WEIGHT MYTH

> A diet is a cure that doesn't work, for a disease that doesn't exist.
>
> —SARA FISHMAN AND JUDY FREESPIRIT

We've been taught that being fat and gaining weight is unhealthy. It's what everyone, *including* your doctor, has been taught. It is our *collective belief system*. We don't really even question it—we just *know* it's true. Fat = unhealthy. But . . . it's just not supported by science. There are so many studies that show that weight and health are not as connected as we have been taught, and that dieting is *not* the cure.[3]

Some incredible research on this subject has been done by Linda Bacon, PhD, author of *Health at Every Size* and *Body Respect.*[4] She has her PhD in physiology, and graduate degrees in psychology and exercise metabolism, and signed a pledge not to accept money from the weight loss, pharmaceutical, or food industry when she got her PhD. Decades ago she began researching weight loss to try and figure out how to successfully lose weight and keep it off, but started noticing that dieting and exercising for weight loss always backfired long-term. After people's initial weight loss, they would gain all their weight back (and more), nearly without fail. Sometimes they would even gain weight when they were still religiously keeping up the diet and exercise regimen that helped them lose weight in the first place. She began to see that our cultural assumptions about the simplicity of weight loss were totally incorrect, so she organized a study to examine this assumption even more deeply.

The Health at Every Size study followed two groups of women in the "obese"* BMI range over the course of two years. The first group I'm going to call the diet group. They followed a standard weight-loss protocol for obesity that focused on a low-calorie diet and lots of exercise. Their protocol was highly regulated and led by one of the top obesity experts in the country. Everything was figured out for them on their plan, and they had extensive check-ins for support to make sure they had everything they needed to stay on track.

The second group I'm going to call the intuitive group. They were *not* told to lose weight, but instead to learn to accept themselves as they were. They started learning to eat instinctively, many of them for the first time after years of dieting. They were

* More on this problematic term later.

taught to listen to their cravings and hunger cues. They were encouraged to enjoy their food and to eat things that made them feel good. They were given permission to move in ways that made them feel good. They were led through exercises in self-forgiveness and self-love, and were guided to heal their shame and guilt over their eating and weight. Essentially, they were taught shame-free intuitive eating.

One of Linda's colleagues was worried that the intuitive group would ruin their health, so she insisted on testing the nondieters' blood lipids and blood pressure three months into the study—and if the markers were getting worse, they'd stop the study. Linda agreed and three months in they were tested, but nothing was wrong with them, so they continued eating what they wanted.

This is what happened over the course of this two-year study: at the beginning, the diet group lost lots of weight *and* their health markers improved, just like we all assumed they would. Calorie restriction leads to weight loss; weight loss leads to better health.

But by the end of the two years, not only had 41 percent of the dieting participants dropped out, but the people who stayed had gained all the weight back—and then some. These women were collectively heavier than they were when they started, even though they were all still dedicated and trying to stick to the diet.

What's even more interesting is that their health markers and self-esteem became *worse* than they had been when they started two years before. For both groups, they were testing blood pressure, total cholesterol, LDL, depression symptoms, and more. All that dieting backfired big-time. And, as you can imagine, they all felt horrible about themselves. So the diet group ended the two years *less* healthy than they started, even though they were still sticking to their plans. Dieting not only made them heavier, but it screwed with their health.

And the intuitive group? The ones who strove to live healthfully and happily as they were? After two years that group had not collectively lost any weight; however, all of their health markers improved. (Again, blood pressure, total cholesterol, LDL, depression symptoms, etc.) They learned to live, move, and eat intuitively, learned to forgive themselves, and start doing activities for the pure joy of it, and they became healthier *without* losing any weight, while still being in the "obese" BMI range. They were able to improve their health without any weight loss at all.

This debunks two deeply entrenched cultural myths: First, it shows that diets don't work long-term. No matter how much support and willpower you have, even if you stick to your diet, there is a biological and metabolic backlash. We believe diets must work because we *initially* lose weight, and *initially* improve health. So when things go south and blow up in our faces, we assume it's our own fault. We don't understand the long-term effects of the diet: the weight regain, how bad it is for our health and metabolism, and the fact that we get into a miserable cycle of self-blame. Really, it's our body's weight-regulation system that is actually running the show all along.

The second myth we can bust is the idea that thin is healthy and fat is unhealthy. These two groups of women show that you cannot tell someone's health from their weight. You cannot tell a person's habits by looking at them. Many fat people are actively on a diet—as they are constantly told to be—and they are trying and failing to lose weight. You *just can't tell* from looking at a person.

Weight is also not as in control of our health as we think. The Health at Every Size movement is asking us to switch our goal from weight loss to healthy, life-affirming habits. Our habits dictate the health we can control, and genetics and other social, emotional, or environmental variables dictate the rest.

Blaming people for their health isn't fair or productive, because healing is *not* easy, cheap, or straightforward. Wouldn't it be nice if health were as simple as eating and exercising a certain way? But it's not. There is no surefire way to avoid illness. Health nuts get cancer and heart attacks all of the time. And doctors and scientists disagree about the healthiest way to eat *all of the time.*

Of course, we want to be healthy. Of course. Wanting to be healthy isn't the problem, but it *does* ignore how much of it is out of our hands. It's ignoring that *right now* at this very moment we are both thriving and dying, and that if we could actually control it, the little 106-year-old Italian woman who smoked and chugged olive oil every day and cited "not marrying again" as the secret to her longevity wouldn't be the centenarian—*we would be.* We would be, and we would credit kombucha and sprouts and be so, so proud of ourselves. But that's not how life works. And it's not how health and longevity work either.

There's even lots of research showing that people in the overweight BMI category live longer than people deemed "normal" weight, *and* that people who are even moderately obese live at least as long as people with "normal" BMIs. Yep. It's true.[5]

Weight-loss studies rarely look at the impact of health and

weight regain *over time*, because it's hard and expensive to do. They usually just focus on immediate, short-term, and temporary weight loss and improvement.

For everyone who is still *sure* that giving up on dieting means giving up on their health, here are some tidbits that will be helpful to hear:

One of the biggest indicators of weight is genetics.[6] We all have "set points," weight *ranges* that the body will try to maintain. No matter how you are eating or moving, there is a weight range your body wants to be in—some people's are higher, and some are lower. Your body will adjust your metabolism in order to keep you in your set point range.[7] We do know that dieting has been seen to *raise* weight set points.[8] Meaning that your body will have a new normal at a *higher* weight than it was before you started dieting. Survival.

Exercise is one of the best things we can do for our health—but in *moderation*.[9] Exercise doesn't *need* to be painful or miserable. In fact, it is way better for you as joyful movement and something that feels good to you—not punishment. Too much forced[10] exercise isn't good for your body or longevity either.[11] Just like dieting, exercise won't necessarily change weight long-term.

Social status and feelings of personal power have more impact on your health than even your health habits.[12] Autonomy and control over your day, your job, your activities, your money, and your life leads to more contentment, which is great for your overall health. And the acute stress that comes from being marginalized or powerless, or feeling shame and prejudice, are all terrible for your health,[13] regardless of weight or *even the way you eat*. The way you are treated, and treat yourself, affects your health.

Not feeling you have any power in your life can make you sicker than any of your health habits[14] . . . that's huge. Experiencing discrimination, *or even just perceived* discrimination,[15] is terrible for your health. And traumatic experiences that are completely out of our control can have major impacts on our health long-term as well. For instance, survivors of the Holocaust concentration camps had significantly higher rates of fibromyalgia,[16] even decades later. And survivors of childhood abuse are at higher risk for having autoimmune disorders.[17]

What this all really means is that we have been blaming ourselves for our health and our weight, when in reality there is so much that is out of our control. And what this also means is that social change, kindness, and empowering ourselves and others will end up being more helpful and important to our collective health than any "war on obesity." There are unhealthy fat people and healthy fat people, unhealthy thin people and healthy thin people. Losing weight does not guarantee you good health, especially if the weight loss happens in a self-punishing way.

The Health at Every Size study is eye-opening and liberating—but it can also freak people out. Because what lots of people hear is, "You mean . . . even if I learn to eat normally, I'm stuck in this body forever!?" What's important to realize is that we *can't* control our weight long-term. We've tried. You've tried. And if you're reading this book, chances are you've consistently failed to keep control—and now here you are.

The good news is, the calmer and more fed your body is, the

better it will work, and the healthier and more stable your weight and appetite will be. Bodies end up right where they belong when you *stop* trying to control weight. The only thing we *can* control is how we treat ourselves, and learning to feed ourselves normally. And the sooner you can accept that your body will handle this whole weight thing for you, the sooner your health and life will improve.

GUESS WHICH INDUSTRY MAKES $60 BILLION A YEAR?

Think about how much money you have spent chasing weight loss. How many books have you bought? How many plans have you subscribed to? How many protein bars? How much money on Fitbits and other weird gadgets? How many pounds of almond flour? How much money have you hemorrhaged for the diet industrial complex? And *what* have you gotten out of it? Chronic low energy, and a deepening distrust in your seemingly insatiable appetite?

The diet industrial complex is made up of weight-loss programs (like Weight Watchers and SlimFast), pharmaceutical and medical companies that make weight-loss drugs, supplements, or procedures, and any other company selling *beauty* and "health." These companies thrive on people believing that they are addicted to food, and that weight loss is the answer to all their problems. And they benefit from all of us feeling insecure, hating our bodies, and believing that we are just five pounds away from becoming the woman we are *meant to beeeeeee*, and at the same time five pounds away in the other direction, from destroying our health.

No matter what they want you to believe, these are *businesses*, not philanthropic charities. They do not care about you. They make *no* promise to do no harm. And these businesses each make hundreds of millions because their products and solutions *don't work long-term*. Because if they did, people would buy one book, or one membership, and become "cured." Then the companies would lose that customer and revenue stream.

It may seem like weight-loss companies sprung up in response to an "obesity epidemic," but when you actually look at the timeline, the opposite is arguably more true. The "obesity epidemic" only came around in the mid-1980s—after people had already been spending decades using cigarettes as appetite suppressants, using amphetamines, ephedra, and Dexatrim, the grapefruit diet of the 1930s, and the cabbage soup diet of the 1950s. Weight Watchers started in the 1960s, and SlimFast came around in the 1970s. But the number of "obese" Americans didn't soar until the 1980s and 1990s, when it doubled among adults in the United States.[18] We all assume it's because of our portion sizes and sedentary lifestyles, but the 1980s and '90s were when exercise became mainstream, and low-fat and diet foods and fake sugar were all the rage. Then low-carb became popular, but "obesity" has continued to rise despite all of our dieting. Do you see how this doesn't entirely add up? Our collective dieting became more and more widespread *first*, and collective weights have only risen *after*, likely because of, and in response to, our dieting and fucked-up eating.

Beauty, health, and weight-loss companies have been telling women what is acceptable and attractive since marketing companies have existed. And we've always been suckers for it. We all want to be beautiful, and of *course* we do when we are taught how important it is for our future happiness, career, love life, personal Instagram lifestyle brand, *whatever*. But diets and body dissatisfaction are also more likely part of the *cause* of rising weight set points, not the cure. Dieting is directly related to people feeling more and more out of control with food.

But companies who sell weight loss have always been seen as the good guys. They want to help us become thin and healthy and happy! Weight Watchers is trying to rebrand because they just want us to live our best lives! Fuck no. They don't care about you. Don't blindly accept that they exist to save us from ourselves. They have always had a vested interest in perpetuating our deep cultural bias against weight, and creating products and programs that only work temporarily so you keep coming back again and again.

A scary truth is that companies that sell weight-loss programs and drugs also have a lot of power at the policy-making level and often fund the studies being used by the medical community. And many weight-loss drug companies sponsor doctors and public health initiatives. One example is our reliance on the bullshit BMI standard.

BMI takes no actual health factors into account. It can't tell you anything about your blood pressure, your glucose levels, your hormones, your metabolism, your strength, your stamina, your bone density, your cholesterol, your immunity, your cellular respiration . . .

nothing. It's literally just a math equation: weight in relation to height, and it was first published by a life insurance company in 1959 as a way of explaining their rates. This was criticized by scientists because the equation it was based on was never meant to be used for individual diagnosis.

But doctors and insurance companies liked the simplicity of the equation, and so the BMI scale became widely used in 1985 by the National Institutes of Health. Then in 1998, the World Health Organization relied on the International Obesity Task Force to create updated BMI recommendations. And at the time, the two biggest funders of the International Obesity Task Force were the pharmaceutical companies that had the only weight-loss drugs on the market. The task force changed the BMI cutoffs on a whim, and overnight millions of Americans switched from being "normal weight" to "overweight."[19] Thanks a *lot*, lobbyists.

The whole thing is arbitrary, because many studies have found that higher BMIs actually have *lower* mortality rates.[20] And many studies have shown that weight loss or too much exercise has been associated with poorer health, higher stress hormones, and increased mortality.[21] And *still*, people are told they're unhealthy based on their BMI, even if their health is otherwise perfectly fine. It's just assumed. *Oh, you're in the overweight category? You* **must** *be unhealthy.*

We can easily compare the diet industrial complex (or "Big Diet") to the military industrial complex, Big Pharma, Big Oil, or Big Tobacco. These are all made up of powerful companies who tend to care way more about profit than anyone's well-being, safety, or the future of the planet, and who have the resources to sway both public opinion and policies that benefit their own interests. In her book *Dispensing with the Truth*, Alicia Mundy calls it "Obesity, Inc." and talks about the million-dollar funding that Weight Watchers and other groups contributed to Shape Up America!, an organization that was part of a strategy to turn obesity into a disease (!!!) so it could be "treated" by the pharmaceutical, diet, and medical industries. That's one reason why I keep putting "obesity" in quotes. It was created by lobbyists.

Our cultural weight bias is so deeply entrenched that even the scientific community isn't immune to it. Bias has the ability to skew the way people interpret and share data; it's called publication bias. Results can be marginalized by the scientific establishment, or even by the researchers themselves, because they don't fit with what is considered to be the truth at the time.[22] Scientists' reputations are at stake when they publish data, and scientists who find results that don't fit with current beliefs have been frozen out of positions, funding, or committees.

Not only that, but most of the studies on weight and obesity that we hear about are ones that are funded by these pharmaceutical and weight-loss companies. Even ones touted by doctors and the government are funded by Big Diet. And when the results don't tell the companies what they want to hear, the companies just ignore the studies altogether.

Drug companies also use tens of millions of dollars to lobby for the approval of drugs that have previously not been approved (because they are dangerous or simply don't work). Drug companies also gave lots of money to medical groups and doctors so they would encourage their patients to use diet drugs.[23] In the UK, the National Obesity Forum was partially sponsored by a number of pharmaceutical companies that *just happened* to manufacture the very drugs that the doctors were suggesting to combat the "obesity epidemic."[24] This is a huge conflict of interest, but this is a consistent phenomenon with big businesses—Big Diet is no exception.

Basically . . . Big Diet is not on your side. It never has been. And not only that, it's all as corrupt as the oil companies back in the 1950s paying off scientists to claim that lead gas wasn't bad for us (hellooooo lead poisoning!), and those cigarette ads kindly teaching us that most doctors smoked Camels.

I'm not sharing this information to depress you—I want to empower you. In order to break free from our fucked-up relationship to food and our bodies, we need to start seeing through the bullshit fed to us. We need to start being our own advocates, in the doctor's office and when people start making hyped-up claims about weight loss and health. Anyone who tries to heal their eating without dealing with the elephant in the room—our own weight stigma *against ourselves*—will not be able to find real freedom and intuition with food. It's all too connected.

THE F-WORD

Let's also talk about the most important and controversial F-word in this book: *fat*. I am going to be using the word *fat*, and I want to explain why. It has become such a loaded word because we've believed that being fat is one of the worst things that we could be. We assume that using the word *fat* is automatically an insult, because people have used it as an insult for such a long time. In the 1800s, even before people had assumptions about fat people's health, fat people were seen as "uncivilized," but were also thought to be *healthier*[25] (probably because many of them were).

These days, one of the reasons that people think being fat has remained an "acceptable" open prejudice is because we think that people's weight is fully their own fault—that their weight *means something* about who they are as a person, and that therefore we get to pass judgment and target them, so we feel better about our own miserable little lives.

Hopefully it goes without saying that whether people's weight is in their control or not, treating a human being poorly because of how they look, or how we perceive their health to be, is cruel. It's never been okay and it never will be, misinformation or not. Fat people are subjected to constant judgment and scrutiny, they get dismissed by doctors, they are passed over for jobs and used as the punch line of jokes. And we all hope that if we can just work really, really hard *not* to be fat, then we can avoid the misery we put them through. *We* can avoid being the punch line of jokes, or being called a fat bitch.

Our relationship with weight, and our deep fear of becoming fat ourselves, is one of the biggest causes of our dysfunction with food. *Neutralizing* the word *fat*, as well as the actual body type, is a really essential step in healing your relationship to food. No matter what we weigh, our fear of being fat is fucking with us all.

There are lots of fat people who are reclaiming the word *fat* for themselves—and unlike words like *curvy* and *chubby*, the word *fat* isn't a euphemism. The word *fat* is allowed to be neutral. That doesn't mean that every fat person wants to be called fat, especially since many people still use the word as an insult, but there is a world where people are self-identifying as fat and trying to take away the stigma of the word and the body type itself.

Words like *obese* and *overweight* are judgmental, medicalized words that were basically made up by Big Diet for profit. So *unless* I'm referring to studies that use BMI directly, I won't use those terms either, and if I do, they'll be in quotes.

All of this being said, I am not fat and I cannot speak for fat people. I recommend you also listen to what fat people have to say about their experiences. But for now, I am going to be using the word *fat* in this book. To paraphrase Hermione Granger, fear of a word just increases fear of the thing itself. I think that applies here.

YOUR DIET MIGHT BE A CULT

Have you ever noticed how fad diets can become cultish? It took me a long time to see the parallel, because I was *in* the cult, and cult members never think they are part of a cult.

Whether you consider yourself religious or not, looking at the parallel between diets and religions, and the societal roles they play, can be very illuminating. For better or worse, depending on your outlook, we are *generally* a more secular culture than we used to be, and in a way, *dieting* is filling a role similar to the one that religions used to fill. For many of us, dieting has become our new religion, and food and weight have become our morality.

Looking at the *positive* side of religion, it offers community, structure, ritual, and an attempt at spreading kindness, love, spirituality, healing, acceptance, and charity.

On the dark side, religions have historically taken advantage of shame and dogma, and ignited our "fear of the other" and people who are different from us. People start feeling like *they* know *the one true way. They* have figured it out. *OUR way is right, THEIR way is wrong. We need to convert the heathens who have yet to see the light and teach them the error of their ways.*

It is the kind of moral superiority that we use to try and make ourselves feel temporarily safe. And through the ages, so many acts in the name of religion have been used as an outlet for the darkest parts of humanity. Witch burning. Holy wars. Refusing to make cakes for people whose personal lives you don't agree with.

So how is this like dieting? Diets seem to offer health, structure, purity, safety, nourishment, nutrition, sometimes environmental responsibility, and—we all hope—a better life.

But diets feed into the exact same human fear that causes holy wars: *I know the way. WE know the way, and you don't. We are doing this right, and you are doing this wrong. We are following the moral and right way to live. This way of living will keep me safe and on the path of righteousness. I need you to hear the good word of coconut oil and follow my coconut oil path.*

I *don't eat grains because I am smart and informed and responsible. I know ALL about phytic acid, and you should too, because YOU are fat and eating all the wrong THINGS.*

We evangelize, we spread the good news, and in a strange way, through diets, we are also seeking salvation and eternal life.

It is our way of convincing ourselves that we are safe. It lets us feel better for a moment because at *least* we're doing better than *them.* It's the dark side of humanity wrapped up in a new cult.

And let me tell you! I have been a member of some *diiiiet cults.* (Mostly through online diet message boards.) I was a disciple! I spread the *word.* I drank the organic probiotic Kool-Aid. I paid the membership fees ($30 for a jar of raw sprouted almond butter). I've been a sucker. I've been judgmental. I thought I was possessed by the devil of refined sugar and food addiction. I've *been there*, and I speak firsthand.

I know what it feels like to *believe.* I know what it feels like to think that your cult is, well, first of all, not a cult. But I know what it feels like to believe that your diet is the *right way.* I know how safe it feels to follow a plan and really, really hope and believe that it will actually deliver on all of its promises.

And it all stems from fear. Fear of the unknown. Fear of mortality. Fear of imperfection. Fear of losing control. Fear of aging. Fear of not being safe. Fear of the sins of the flesh. It's sad, it's lonely, it's isolating, and it is so, so human.

Part of the big problem with the diet and beauty industries (and many other industries that capitalize on your insecurities) is that your fears are being exploited. They want you to believe you aren't good enough as you are. They make you believe we are all supposed to look the same. They want you to believe that you need them to save you.

So if there's any part of you that's looking at *me* and hoping you end up where I ended up, or looking at anyone else and hoping to end up where they ended up, that's a habit that I want you to become aware of. It's a very human habit, we all do it, but it's not helping. Trying to be someone *else* is what got us into this mess in the first place.

Your best self is probably the one who trusts yourself the most, is able to relax and be social when you feel like it, and is able to seek quiet time when you need it. Someone who is able to be spontaneous when it suits you, and willing to take up space, speak up, take risks, use your creativity, is willing for things to be messy and imperfect—and is an all-around happier human.

Some people are hesitant to go on The Fuck It Diet because they don't know if they actually like who they really are. They're not sure if who they really are is that special or interesting or attractive enough. I get it. That's a scary thing to think. Thanks to lots of insidious messages from the media, from princess fairy tales, from family, from dysfunctional relationships, from other insecure women, or from diet, drug, fashion, and beauty companies, it can be hard to trust that you're actually okay as you are, and that you don't need to change or appease anyone. We'll be exploring these concepts even more later in the book.

I want you to free yourself from diet cults, but I'm not ragging on God. I am a *big fan* of spirituality and "whatever word you'd like to use for God." But beware of dogma. You can tell it's all going south when you are experiencing lots of fear, judgment, and feel all-holier-than-thou.

Here is where I also tell you that once anyone starts making The Fuck It Diet into a cult—including hypothetical, foolish future-me—that is when you remember that you are your *own boss*, and that your own intuition is king.

DIETS UNRAVELING

Right before my own Fuck It Diet epiphany, I was paleo and kicking myself for eating too many bananas. It was around the holidays, and I was bingeing daily on paleo ginger snaps and paleo pumpkin pie made out of butternut squash and honey.

This had been my pattern for ten years. I would follow a diet religiously for a month or two or ten, and eventually find myself constantly hungry and thinking about food. Then I'd start to really take advantage of the "allowed" foods, normally bingeing on them at midnight. I would be furious with myself and every morning would try to regain control. Eventually I'd stop the diet completely, heartbroken that it didn't heal me, or my bingeing, or my food addiction, and move on to another diet.

And now, here I was *again*, gaining weight *again* because I couldn't even stick to a reasonable, very-low-carb paleolithic diet like the one our ancestors apparently ate. *Get it together, Caroline!*

My first inkling that something might be truly wrong, beyond my self-diagnosed "food addiction," was when I started walking by the mirror and having really opposite reactions just a few minutes apart. I'd walk by and think, *WOAH, I'm actually really thin . . . weird. I guess I didn't gain ten pounds from all the almond flour ginger snaps I ate in bed last night.*

Then a few minutes later I'd walk by the same mirror and think, *WHAT!? How am I so big!? Oh GOD! Look at my FACE!* Then the next morning, *Wait, wait, I actually do look thin. WTF.* I felt crazy.

It was only a month later when I had what I refer to as "my epiphany." I was staring in the mirror over my bathroom sink, and it hit me like a bolt of understanding. I realized that my dysfunction with food was never going to change if I kept getting into this cycle over and over again. It would never change if I held on to my need to be skinny. In one moment it became so clear to me that not only was dieting metabolically backfiring, but my relationship to my *weight* was the core cause of my misery.

What came after the epiphany was hard, but the decision in that moment was simple. I intuitively believed that if I could surrender to the process, it would all work itself out—mind, body, and spirit. Nobody could promise me that it *would* work out, but on a deep level I knew that if I could be brave and embrace a higher weight, and feed my body what it needed, then I'd be free.

WHAT IF YOU'VE ALREADY TRIED?

Most of the people I work with have already tried to heal their eating. They've tried intuitive eating or some other version of "just be balanced" or "just listen to your body." They come to The Fuck It Diet after being so frustrated that they Google "Why doesn't intuitive eating work?!" Really. That's the number-one search phrase that brings people to my site.

If you've tried to heal your eating by not dieting before, and it didn't work, that is most likely because you were ignoring your relationship to your weight and still trying to make intuitive eating into some kind of diet. Most of us think that if we can just "eat intuitively," we will eat like a bird and become the naturally thin and happy version of ourselves. So many of us try to heal our eating without changing our relationship to *weight* as well. Ignoring how closely our feelings about eating and weight relate to each other is our big mistake.

Before my last-ditch-effort diet on paleo that led to The Fuck It Diet, I thought I was "eating intuitively" *for six years*. I thought that intuitive eating was the same as "sensible portion control." I thought my "successful" stint of trying to eat like a "French woman" was intuitive eating. But it's all a fucking diet in sheep's clothing.

Now I realize that the entire time I thought I was eating intuitively, I was still focused on weight, and still scared of most foods, whether I was letting myself eat them or not. My intuitive eating was still used to try to eat *less*, which is inherently going to backfire.

YOU'RE IN CHARGE (FINALLY)

Think of all of the unspoken things that dieting promises: that if you follow this simple four-month plan, you will become someone else—someone better. Eat only raw foods and practice daily sungazing at dawn, and not only will you be beautiful, but you will transcend this earthly plane. The promise is that with lots of willpower, you can obtain a perfect body, and when you do you can finally be proud. If you follow someone else's rules, everything will finally become perfect and easy. And if you let yourself slip and gain weight, you should be ashamed.

Obviously, all of this is a recipe for physical, mental, emotional, and existential disaster.

The Fuck It Diet promises none of those things. You will

probably not obtain your old definition of a perfect body. But you *will* get your calmest, happiest body, without the extra stress and yo-yo and impaired metabolism. And to get there, you can't follow anyone's rules but your own. Not even my rules, because *my* whole goal is to get you to a place where you are able to trust and follow *your* impulses and intuition and appetite, without the absurd pressure of weight control and weight loss.

Before diets—even if you can't remember it—there was a time when you knew how to eat, and you didn't see yourself or your worth based on weight or food, even if that was all the way back when you were a little kid.

This is no longer a journey of control, willpower, and perfection. This is a journey back to whoever you were before diets, before you veered away from yourself and went down a path that took you here, reading this book. That diet path was a path of listening to what other people expected and wanted of you, and the never-ending saga of trying desperately to get approval from anyone and everyone but yourself. You can keep trying to grasp onto that control, but it will continue to be the miserable, tragic, exhausting saga that it has already been.

This book is going to encourage you to unlearn all of the things that made you stop trusting yourself. And you will have to relearn all of the things that will *allow* you to trust yourself again. What this also means is that your specific journey is going to look different from the next person's.

It's important to say that this book is not a quick fix. The Fuck It Diet is basically a life-and-heart overhaul. This isn't a thirty-day Fat-Burning Extravaganza and "now you're happy and beautiful forever" kind of thing. It's not a "this new shiny-*and*-matte lipstick will never come off and you'll be beautiful and impressive and happy all weekend" kind of thing. It will probably be really scary, because I'm asking you to let go of so many of the things that used to make you feel safe and worthy. Instead, I want to help you find ways to feel worthy that *transcend* the way you look or how impressive you convince yourself and other people you are.

The rest of this book will be helping to heal your relationship to food *and* weight: How to actually live a life without diets.

There are four parts that make up The Fuck It Diet journey, so to speak: physical, emotional, mental, and then the final thriving part once you have your life back. Fair warning: because this is a book, I had to choose an order, but these steps are not really linear at all. I wish for both of our sakes that they *were*, because it would make going on The Fuck It Diet easier. It may be helpful to plan on reading this book twice. The first time through, just take in the different things you may experience, and then the second time, slowly address all the areas yourself and apply the exercises more deeply. But again, you're the boss. You do what feels right.

THE PHYSICAL PART

In this section, we are going to reverse physical restriction and its biological effects by eating. This is the part that usually shifts the quickest. It's not as difficult or complicated as you might think to take the body out of crisis mode. All it takes is a lot of food and rest. Luckily for you, getting out of the biological starvation mode will directly help a good chunk of your obsessive mental fixation on food, too.

THE EMOTIONAL PART

Next we are going to talk about our emotions, and how important it is to come back into our bodies and feel what's waiting to be felt. We will address emotional eating, how it is different from bingeing, and how to address it *without* restricting. We will also be talking about another survival state in this section: our old friend fight-or-flight, and how this state is directly tied to old unresolved emotions. Becoming more willing to feel all of the things you have avoided and pushed down through the years will help not only your relationship with food, but everything else in your life too.

THE MENTAL PART

Whether we know it or not, we have absorbed so many rules about eating, food, and weight that don't serve us. These rules become our beliefs, which are able to affect everything we do and think and feel. Our beliefs have a lot of power over us, especially if they are lurking in the shadows. So in this section you are going to learn how to become *aware* of your beliefs, and learn tools to lessen their power over you so you can begin to see clearly again. Whew!

THE THRIVING PART

My ultimate goal is to get you fully intuitive with food. Once you get out of your own way when it comes to eating, this is where the fun stuff happens. In this last section we focus more on deep rest, self-care, boundaries, figuring out what you actually *enjoy*, and more. This is where you get to really discover who you want to be without the distraction of food and weight.

Throughout this book I will also be sharing five main tools that will act as your *anchors* on The Fuck It Diet. They are all very, very simple, and will not take that much time, but they will make a huge difference. Don't let their simplicity fool you; they are game changers. These are the five tools that I hope you take with you into your life after you finish this book.

But! None of this healing happens by *thinking* about eating or *thinking* about giving up control over food—it only happens by *doing it*.

And with that being said, let's do it.

TWO

HOW THE HELL DO I ACTUALLY DO THIS?

THE PHYSICAL PART

How to go on The Fuck It Diet:

1. Stop restricting.
2. Trust your body, appetite, and cravings.
3. Eat deliciously and normally for the rest of your life.
4. Embrace life in a (probably) *not*-stick-figure body.
5. Do cool, fun things, and enjoy your life.

If you easily did those five things, I wouldn't need to write the rest of this book. But we are so petrified of and resistant to this process. We have spent sometimes *decades* assuming that the only way to be happy and healthy is to micromanage our food and our weight, so this entire thing gets really emotionally complicated, really fast. We get stuck in old patterns and triggered by old fears and need a lot of help getting through all that. We have a lot to unlearn.

We are going to start with the most concrete part of this journey. This is where you *physically* heal the body from the restriction and famine state. And the way to do that is very simple: if you are hungry, eat. Yep. That's it. It *is* actually that simple. But what's so hard about this is that we have *so* much resistance to eating as much as we are hungry for. Sometimes the simple things are the hardest, especially when we have to relearn to get out of our own way.

Obviously, people's eating needs will vary based on their own bodies, and how much and how long they've been restricting. But even people who don't think they've been restricting much may end up being hungry for way more food than they think they will. The beautiful thing is that you don't need to know how much that is, you just need to follow your cravings and hunger.

What's great about this phase is that it's very straightforward. You will hold the cure in your hands, and then . . . eat it. We need to neutralize our relationship to food, not just by *thinking about it*, but by eating it. You have to give your body what it needs: real nourishment and rest. And if this sounds like crazy and irresponsible advice, think carefully about what I just said . . . Being told to feed yourself and listen to your hunger and trust your body is radical? What?

In this physical part, progress will be tangible and concrete, and you will quickly experience the benefits of eating and rest.

» TOOL #1: ALLOW FOOD

Yes, this is how basic we are getting to start. Allowing and eating food is the most fundamental building block of this entire book, and is the most fundamental building block in life, too. You need to *allow* all food, whenever you are hungry, until what is guiding your food intake is your body, intuition, and desire, and *not* the famine response and your scared, hungry mind.

I know that "just eat" sounds counterintuitive to everything your mindfulness teacher told you, but I am a mindfulness teacher too, and I am telling you that if you are hungry or underfed, you can't be mindful or peaceful at all. This is a foundational concept in psychology called Maslow's hierarchy of needs. Food and rest are two of humanity's most *basic needs*, and if those basic survival needs are not being met, it is almost impossible for us to truly move on to any other area of our life.[26]

You *need* to eat, multiple times a day, every day, for the rest of your life. If you're hungry, it means you need to eat. You're hungry but you just had a snack? Great! That means you need another one. If you are hungry, eat. No matter how much you've already eaten. That's what hunger is for. That's how it goes to be alive and human.

Also, when I say to eat, I am not just addressing obviously underfed or emaciated people. I am referring to *everyone* who is fixated on food, no matter how much you binge, and no matter how much you weigh. This tool applies to you no matter what your body size is. Yes, really. *And* it applies to you no matter whether you tend toward the more restrictive side or feel like you tend to be more of an overeater/binger. Bingeing is a natural reaction to that famine response. So, making sure you are eating to nourish yourself is the only answer—always.

Eating is healing to both the biological and psychological

sides and is the *only* way to heal each of those sides. Your metabolism is impaired by restricting, and the only cure is eating and gaining weight.

Any time you eat a lot, remember that it isn't your body losing control—your body is doing all of this on purpose. Eat. It really is that simple.

DISCLAIMER: This book cannot begin to account for or take responsibility for everyone's individual health needs. For instance, if you are diabetic, celiac, or have serious food allergies, you obviously need to eat within those parameters. And if you don't feel well, go to a doctor. I recommend a doctor or nutritionist who is well versed in Health at Every Size and nondiet approaches to health, who can support your health from a weight-neutral place. Doctors work for you, not the other way around, so find someone who will help support your health while you learn to have a better relationship with food and weight.

THE NOBLE ROLE OF WEIGHT

I know what you're thinking . . . *I think I can try to do this without eating too much and without gaining weight.* Get that idea out of your brain. I've got your number.

Resisting gaining weight is going to keep you stuck. It keeps so many people stuck. I know you wish I could tell you *exactly* what will happen with your weight, but I can't. I can't predict what your weight journey will be. But I *can* say that your particular weight trajectory will *probably* look similar to your *past* weight trajectory, because we all have weight set point ranges. Weight set range is controlled by our hypothalamus, which regulates our eating and activity habits and metabolic efficiency to keep us at the weight where our body feels safest and healthiest. Weight set points are diverse in any population and are not really within our control.[27]

It's never been your fault, and there isn't much you can do but take care of yourself . . . and eat. In the beginning of this whole process, I gained weight, but not as much or as fast as I assumed, and at a certain point it just . . . stopped. I kept eating, but my weight stopped going up—I'd reached the top of my weight set point range. And as my eating and metabolism continued to normalize, my weight went *slightly* down. Then about a year in, I reached a point where it didn't really matter what I did or ate, whether I moved or not—my weight stayed the same, and it has stayed the same for seven years since, except for subtle natural seasonal and hormonal fluctuations. It completely takes care of itself. Life-changing.

Most people who go on The Fuck It Diet gain to the top of their weight set range, and then eventually, slowly, without trying, fall to the middle of their weight set range and stabilize there. Meaning, if you have yo-yoed about thirty pounds in the past, doing The Fuck It Diet will probably keep you in a similar range. Same if you yo-yoed eighty. I know that may be a tough pill to swallow. But the important thing to remember is that *dieting* is the thing that will continually force your weight set point up and up and up. It is the act of trying to control your weight that encourages your body to put on more and more and more weight. So, if you need to use your fear of a rising set point as a way to encourage you to stop dieting, fine. Use it.

And all of those times when people naturally fluctuate and gain weight, for instance, in the winter, instead of panicking and fighting it, it's actually safer to just . . . let it be. Our tendency is to assume our weight is a sign that something is terribly wrong, and that it is going to go up and up, and so we immediately try to diet to get the weight off—but the body knows what it's doing.

Paradox of paradoxes, allowing yourself to gain weight is the only way to stop this cycle. And I don't just mean allowing yourself to gain *whatever amount you have decided is an acceptable amount to gain.* I mean . . . really surrender. *Fuck it.* Choose to listen to your body. Throw your scale out the window. Flush your Fitbit down the toilet (actually don't—use the garbage disposal). You will normalize faster than you think. The faster you trust, the faster you will stabilize.

This isn't hocus-pocus . . . this is biologically sound. The body's genetic weight set range is affected only slightly by lifestyle—including dieting, which, again, can push the weight set range higher.[28]

Basically, ultimate long-term control over your weight is an illusion. The times when you felt you had control were actually the times you were setting yourself up for the backlash (the binge and regain). This is the way we are wired, this is the way we have been for many, many thousands of years, and your attempts to override this biology will be continually futile.

Not only is weight gain an important part of fixing the metabolism, but it is also an emotional rite of passage. We all need to face our fears of being a higher weight. We need to learn to be happy and fulfilled at that weight. We need to learn to accept ourselves there. We need to let go of the fear of what "gaining weight" means. We need to be willing to dress ourselves at a higher weight. We need to learn to value ourselves at any weight. It is essential.

I want you to imagine that you went on The Fuck It Diet and never gained any weight. Though that might seem ideal, it is not, because you would continue to live in fear of what would happen to you if and when you *did* gain weight (which you inevitably will with illness, pregnancy, menopause, broken ankles, etc. You are not a robot). You would continue to live in fear of gaining weight. You'd keep living in fear of what people would think of you, how people would treat you, and what you would think of yourself if you gained weight. And all of that subconscious energy would affect your eating and the way you experience your entire life and body.

I have worked with so many people who *first* hoped that TFID would not affect their weight, but eventually found that clinging to that hope halted the process. But when they eventually let their weight do what it needed to do, it made all the difference and led to a way bigger sense of freedom. Not only because they faced their fear and *actually* let their appetite heal, but also because they experienced firsthand that it was not exponential weight gain. There was a natural place where their weight stopped and their appetite normalized. This is true for people all over the spectrum as well, for both naturally thinner and fatter students: accepting gaining weight during this process is both crucial and possible and makes all the difference.

One of my students shared with me, "I never thought I would

say this, but I am so grateful for weight gain. Since TFID I have gained weight and have changed shape. Now that I have, it flabbergasts me that I was only able to accept myself when I was thinner (not surprisingly, I wasn't actually accepting myself at all . . .). I am so grateful that I get to challenge that limitation, and truly find out that I am fine at whatever size. It has really made all the difference in my happiness."

I know we are petrified of what other people think about our weight. But even people who make comments on other peoples' weight don't *truly* care about your weight. They are thinking about their own. What matters more is what you think about *your own* weight. And luckily that is something you can change.

You must accept gaining weight in order to end up enjoying and experiencing all the emotional freedom of The Fuck It Diet. There are no two ways around it. And if you can't jump on board with accepting your body yet, at least accept that you are going to have to accept your body. That's a good solid start.

Throughout this book, I am going to share some writing prompts and activities to help you take these concepts off this page and into your life. Don't overthink them, let the writing exercises be casual.

CONSIDERING WEIGHT NEUTRALITY

Make a list of at least five reasons it may not matter what you weigh.

For example, "My aunt has always been fat, and she is everyone's favorite and a famous painter." "My health was better before I started dieting." "My weight was not a question on the bar exam, so it must not affect my ability to do my job." "The happiest relationship of my life was when I was fatter." These are just examples. My aunt is not really a famous painter.

You can take from your real life or from this book. Go for even more than five if you can.

YOU ARE NOT A CAR

You are not a robot or a machine. Your fueling system *does not* work like a machine. You are significantly more complicated than that, and your metabolism has evolved to slow down when you are not consuming enough. That means that calorie calculations are rendered totally moot and pointless when you understand that your metabolism is adjusting itself to *purposely* keep on weight when it senses restriction.

Your body is trying to conserve energy and also trying to get you to eat. So, restriction of any kind will cause you to fixate on food, be hungrier, more tired, and put on weight quicker—all to save your life. Those symptoms are often signs of a slow metabolism.

Even though you *are not a car*, I am going to use a car metaphor anyway: Imagine that eating is like "revving" the metabolic engine. Stoking the fire. When you eat, you are increasing the metabolic fire. You are allowing and *encouraging* the metabolism to speed up to digest and process the food eaten. The more you eat, the more you encourage your metabolism to work. The less you eat, the more it slows down to conserve and save your life.

Eating and resting is what the body is asking for, and indulging it teaches the body that there is no more restriction. It allows the body to put on weight just in case of another famine/diet and encourages the body that it is safe to speed back up to a normal metabolism and slowly get out of conservation mode.

If you re-feed yourself after a famine, eventually your weight will stabilize in a range that's right for you, your appetite will normalize and become easier to satiate. You'll also have more real energy, more willingness to go for a walk, and more of a craving to move your body, not out of stress or worrying that you're getting flabby, but because you genuinely just *want* to move.

Chronic exhaustion is something many dieters experience, having no idea that it's from their *attempts* at healthy eating. One of my students, Diana, said, "I HATED exercising. It was a tedious chore. And I'd been to several specialists over the years in an effort to find out why I was always so tired and so low energy. Nobody knew. Turns out, I just needed to eat—a lot!"

Another reason why calorie calculations are moot is that different bodies will be able to extract different amounts of calories from what is eaten, depending on the health of their digestion. The healthier your digestion is, the better you will be able to utilize the calories you've eaten. The *worse* your digestion is, the *less* calories you will be able to assimilate from your food. To a diet brain, having less calories able to be digested may sound like a good thing. But *no*, worse digestion, and less calories absorbed, is *not* better for your health, because "less calories" is not better.

The most important thing for me to remind anyone who is worrying over whether their metabolism is healed or not is that no matter whether it *is* or not yet, eating is the answer. Eating and resting will heal a stressed or repressed metabolism. And eating will keep a healthy metabolism healthy. If you are hungry: eat. It's healthy to be hungry. Food will both sustain and repair all your organs and muscles.

If you are coming from a place of restriction and metabolic suppression, eating is *the only thing* that will allow the metabolism to speed up to normal, healthy speeds again.

FAMINE LOGIC

If you'd been in a famine, how long do you think it would take to re-feed yourself? A few months? A year? What feels right based on how long you had been restricting? And when would you know things were back to normal?

OUR FEAR OF HUNGER

Do you eat preemptively to make sure you never have to experience hunger? Do you fear feeling the pangs because you know that a binge might be waiting? Do you feel like you do not deserve to eat unless you're absolutely starving? Do you panic when you are still hungry after you finish a meal? Do you panic when you are full after you finish a meal? Do you anxiously stuff yourself when you're hungry because you fear that if you don't get it all in now, you'll never get to eat again? All of these are ways that our fears and habits get in the way of normal eating.

When I was a freshman in college at NYU, I was a raw vegan. I was also in school in a top program for musical theater, which had always compounded my weight obsession. But what outshone my identity as an actor was my identity as an obsessive, neurotic raw vegan who was spending exorbitant amounts of money on raw vegan dehydrated food and cashew cream desserts.

I would spend my one-hour lunch break traveling twenty minutes to, and twenty minutes from, a raw vegan health food store in the West Village. Then I'd spend the remaining twenty minutes shoving an impossible-to-digest salad and weird sprouted nut dessert into my poor, struggling body before I went in to the afternoon acting class.

I'd spend acting class stuffed with sprouts and cabbage that refused to digest, completely self-conscious about my horrible skin. *Why isn't raw veganism making me GLOW like they said it would?!?!?!*

I put my body through intense and insane diets in the name

of health, but raw veganism was probably the most intense and insane one of them all. That Christmas I brought a papaya to Christmas dinner . . . and that was all I ate. When family members and friends asked me how long I planned to eat this way, I genuinely replied, "Forever." It wasn't going particularly well, but I had to buy in. I had to convince myself it was. I was also convinced that it was just a matter of time before it paid off and healed me.

A month later, I said I was sick one day in ballet, but I was actually just on a juice cleanse. And because I was already a raw vegan, this was like a cleanse *within* a cleanse, a no-sugar juice cleanse, which meant I was just drinking $12 cucumber and kale juice for a few days. I was freezing and weak, so I got to sit down against the mirror and watch everyone do their pliés. I didn't feel guilty, though, because I was convinced I was just a few days away from detoxing *all of my health issues.*

I was never officially diagnosed with any eating disorder. I was never thin enough to concern people, and I was able to hide it from *myself* under the guise of just being a "health nut." This does not mean my eating wasn't disordered. But according to our cultural parameters for an eating disorder, I didn't *quite* fit the bill. What's the medical term for "all I fucking think about is food and dieting and weight and toxins"?

Thanks to biology, my body tried to make up for the missed calories in the days after the cleanses, which is what the cleanse gurus expressly tell you *not* to do. They say things like, "Only eat oranges for a few days after your juice cleanse." I'm sorry. WHAT? To me, and most other people, there is absolutely no getting around a binge. The body's biological response to famine *is bingeing.* And if you try to fight it and actually succeed? That's what we call anorexia, or a very active restrictive eating disorder, because you are able to override your biology. But this is also why I never thought I had an eating disorder. I couldn't fucking *stop eating.*

Our bodies are wired to be afraid of hunger, hormonally. The hunger hormone ghrelin rises whenever you haven't eaten enough food, and it makes you extra hungry *and* slows down your metabolism to conserve energy until you eat enough food. Your body is trying to guard you against eating less and dying, so it wants you to fixate on eating as much as you can get your hands on. It *needs* you to try and avoid long periods of hunger. That's how a species survives—prioritizing eating and satiation.

Hunger begins to feel like the enemy in so many ways. And, not for nothing, diet books and gurus also make it seem like hunger is an issue that we need to eradicate. *Hungry all the time? It's because you're eating the wrong foods! Eat the foods on my scientific plan, and you'll literally never be hungry again!*

So now you associate hunger not *only* with biological discomfort and panic, but also with bingeing and feeling like a total failure.

Let's take a moment to remember that people who are not hungry are usually very sick. Lack of hunger is *not* a good sign. It means that something is wrong, and maybe you are dying. Still, I cannot tell you how many diets I have read and followed where the promise was always some version of becoming cravingless and not hungry. The message was always that your *hunger* was sabotaging your attempts at health and beauty. This creates a major disconnect between you and your body. I mean, if you can't trust your body's signals, then what *can* you trust?

If we are trying to get to a place where we are cravingless, we may as well snort cocaine so we're never hungry OR tired. Then some heroin so we never have to feel. Screw being human. Diets and drugs. Numb all your healthy, normal human bodily functions. What are we *doing*?!

Getting out of famine mode is the closest to "not hungry" you will ever want to get. Basically, get your body to a place where it's not *scared* that it won't get to eat. You'll still get hungry, it just won't feel so epic and out of control. Now I know that when I get hungry, I have the liberty and obligation to eat as much as I want and need.

If you have any sort of fear of hunger, you can cure it through consistent, ample eating. Surprise, surprise!

The more eating is expected, allowed, consistent, and rule-free, the more your body and mind will be allowed to calm down about it. The more you consistently let yourself eat, the more you will learn that hunger is just a completely normal and fixable part of the day.

When you are hungry you are supposed to, and allowed to, eat. You are also allowed to get full. And when you finish your food, and are still hungry, you didn't eat enough. It's that simple.

And when you stuff yourself with food because of your past fears of not being allowed to eat, that's okay too. It's all learning. I know it sounds too simple to be true, but your compulsion to eat more than you actually want is a symptom of the old rules and fears that there won't be enough food, or that another diet is coming. So make your body sure that another diet is not coming. Hunger is not a failing on your end. The goal is not to eradicate hunger. The goal is to befriend it.

After years of believing that hunger was my actual mortal enemy, I am now hunger's good friend. Though, in a way, hunger *is* a mortal enemy. It'll actually kill you if you ignore it. So . . . stop ignoring it.

YOUR RELATIONSHIP TO HUNGER

What is my relationship to hunger? What do I think, fear, wish, judge about hunger? What do I believe about hunger? How do I try to manipulate hunger? Write anything that comes to mind.

THE DIET PENDULUM SWING

For every action, there is an equal and opposite reaction. That's freaking *science*. And that is also exactly how a diet and its aftermath work. The pendulum has to swing: you will be hungry for lots of food. There is just *no* way to go straight from restricting to absolutely normal, easy breezy eating. It is impossible. Not only can your brain not quickly make that switch, but your body actually needs "excess" food in order to heal and repair and normalize.

You are almost definitely going to be really hungry for a bit. You are most likely going to want and need more food (for a while) than you ever thought was acceptable or healthy. And you are going to have to embrace it fully.

You may think you've eaten more than enough and be shocked and annoyed that you are hungry *again*. You may be stuffed but not sure whether you are actually full yet. This will freak you out. But this is normal. And it *will* pass. The pendulum will swing back naturally.

Lemme be clear: the goal of The Fuck It Diet is not to heal your appetite so you stop wanting food. Let me repeat: *not being hungry is not the goal.* Delicious food is allowed to be a part of every day, for the rest of your life.

But remember, allowing the pendulum swing of hunger and cravings will allow you to get to the place where you don't need to overthink food. Embrace this pendulum swing. The more you resist it, the more miserable you are going to make yourself, and the more stagnant this process becomes. If you resist your hunger and your cravings, it's gonna halt it and lock it, instead of just letting it run its course. Resisting your hunger is not the answer. Embracing it *is* the answer.

The fear we have with embracing and allowing our hunger goes something a little bit like: *If I give in to my hunger, it will take over my life, and it will never end, and I will eat the whole world, and then I'll explode, and then people will roll their eyes at my funeral.*

You are not going to eat the whole world. You are not going to eat until you explode. You are not a bottomless pit. And you are also not the one person this won't work for. It doesn't matter if you've been a food addict your entire life—the reason you are is *because* of restriction. And the cure is the complete removal of restriction.

EAT WHEN YOU'RE HUNGRY

Another one of the most searched phrases that leads people to my site is "I'm so hungry what do I do?" Ummmm . . . What do you think? The fact that we are all wondering how to get rid of hunger instead of just eating is *insane*. At this point, so many of us are *so confused* that we now think hunger is some kind of horrible problem that we need to heal with anything but food. But the answer is not tricking your brain by eating on a smaller plate, or filling up on water or caffeine, or trying the newest appetite suppressant herb. *Eat*. The answer is, *eat*.

Diets have taught us to steel ourselves against hunger. But I am telling you the obvious opposite truth: you can and should *always* eat when you are hungry. Beyond it just being biological common sense, you need to prove it to yourself—for a long time—that you will answer the important biological call of hunger. You cannot expect your body and nerves to calm down until you've been amply fed for a while.

If you are hungry and you want to eat, you should eat. Even if you think you have already eaten too much. It really is that simple. Does that mean that the SECOND you get hungry you *must* eat? No. You don't *have* to do anything. But in the beginning, you will need to prove to yourself that you are willing to feed yourself no matter what. Down the road, you will *know and trust* that you will feed yourself, so waiting until it's more convenient to eat will be a nonissue.

Hunger is not a sign that anything is wrong, it's a sign we need food. A reader reached out to me and said, "My colleague just said to me, 'I have already had two liters of water, *why* am I still hungry?' I told her she needed to eat and she looked at me like I had two heads. And to make matters worse, we are health-care professionals. . . ."

There is widespread confusion about hunger and health. Thinking of hunger as some kind of problem is really toxic to our relationship with food and our bodies. So if you have lingering beliefs about times of day you should or shouldn't eat, any rules about eating right when you get up, or waiting to eat when you get up, or waiting a few hours between eating and going to bed . . . all these rules can go jump off a bridge. I want you to become super aware of them, and any other beliefs about when and how you should be eating.

I spent the first three years of The Fuck It Diet eating a *lot of food* in bed every night at midnight. Because for the life of me, I didn't know how to change my circadian hunger rhythm to be hungry earlier . . . if that was even the issue. So I just surrendered, and trusted. I ate a lot when I was hungry, which happened to be at my bedtime. And look how amazing I turned out. (Really amazing.) After a few years it's totally shifted—I don't need to eat much at bedtime anymore, and sometimes I'm not hungry at all—but it took a long time and a lot of trusting before things shifted on their own.

I am not telling you to do what I did. I am just trying to illustrate that you can do the thing that everyone tells you not to do, and it can be a part of the thing that improves your life 1000x. Everybody else's rules mean nothing.

I want to stamp out *any* lingering guilt or stress or overthinking about food because it is—and will always be—completely unhelpful and pointless, and will only lead to reactionary eating. You've been trying to control your natural hunger for years, and it is time to feed yourself.

Intuitive and mindful eating gurus will insist you *only* eat when you are hungry. And while I agree that it's almost always more *pleasant* to eat when you're actually hungry, "Only eat when you are hungry" is *not* a rule on The Fuck It Diet.

Do you know what happens when you believe you aren't allowed to eat when you're not hungry? You'll start overthinking hunger and whether you're "allowed" to eat. That causes confusion, stress, denial and . . . eating when you're not hungry. The Big Paradox. Then you'll feel *guilt* because you weren't "allowed" to eat, and you're in the vicious cycle again.

Eating when you're not hungry isn't a crime, or even a problem. And to a normal eater who feels neutral and easy with food, eating when you're not hungry will happen. Your coworker makes cookies and offers you one right after you finished lunch. Are you hungry? No. Do you eat one anyway? Yes! Sure. *It is no biggie.*

Eating birthday cake or another special dessert, or trying someone's food when you're finished with yours, eating more of your dinner than you needed, or eating something to tide you over when you know you won't be eating dinner until later—those are all *normal* ways of eating. And those are times of eating when you aren't necessarily hungry.

Our goal with The Fuck It Diet—unlike *diet* diets—is to neutralize all food situations that used to overwhelm you. The less food is a big deal, and the less there is a way to "mess up," the less you'll overthink and drive yourself into reactionary eating. When we are fed and food is neutral, eating when we aren't hungry isn't really even that great. It's kinda dull and uncomfortable. And the more it's allowed, the less exciting it'll be.

The truth is, eating when you are full is most likely something you *won't want* to do much once you're more neutral with food. But again, true intuitive eating does not happen by rating your hunger on a hunger scale. It happens by eating impulsively and instinctually when you're hungry, and easily stopping when you're full. This is not something you can force; you need to let this process of trust and intuition happen naturally, *by eating*. But until that naturally happens, you need to let yourself off the hook if you eat when you're *not* hungry.

People tell me, "But my issue is that I eat when I'm not hungry *all the time*." For people who feel like that is their biggest issue, eating when they're not hungry is a nervous response to *something*. And most often and most importantly, it's a nervous response to restriction. It's fearing you won't have access to whatever food you want. It's fearing that you won't let yourself eat the things you want or need. It's a nervous response to restriction or the fear of an impending diet. There is a fear that there is another diet coming. It's a response to the subconscious fear that you're going to end up very hungry at some point, so you may as well get it all down now while you're not even hungry.

So what do we need to do before anything else?? *ALLOW ALL FOOD. ALL THE TIME. EVEN WHEN YOU'RE NOT HUNGRY.*

Yes, I know. It sounds nuts. I don't care. It is the only thing that works. It's the only way to become normal and easy with food.

I know you're thinking, *BUT WHAT ABOUT EMOTIONAL EATING?!?!* Yes, humans *can* eat compulsively to avoid feeling lots of other things, and don't worry, we're going to cover that in the emotional section of this book. But remember, you can't help or cure "emotional eating" by trying to control it—that's restriction. You have to deal with the source of pain, not the symptom, which is what we are going to do. Again, we will be doing lots of work on feeling and emotions soon—I promise! It is coming! But trying to stop your eating, even your "emotional" eating, is going to put you back in a restrictive cycle. Restricting is not going to help with emotional eating, ever. Getting rid of restriction *has* to come first.

Fear of restriction can cause compulsive and emotional eating. As long as there is still a background fear of being denied food, you will compulsively eat. It doesn't matter how many of your emotions you feel and process. It doesn't matter how mindful you become. It is impossible to isolate or help the other causes of compulsive eating when you are also eating because of your biological and emotional subconscious fear that there is going to be another diet. The answer is proving to your body and mind that there is

NO famine and no upcoming diet. And the way you do that is by eating.

I know, I know, the other big fear here is that you'll train yourself to eat when you aren't hungry, and *keep* eating when you're not hungry, until you die in your bed at 1,240 lbs. But it just doesn't work that way. What you've actually *already* done is trained yourself to eat when you aren't hungry *because* of your subconscious fear that there won't be easily accessible food for you to eat. You've already accidentally trained yourself to stay in a constant dysfunctional hell with food. Eating is the way out.

Janet wrote to me saying, "Last week I experienced food neutrality I had never experienced before. My friends wanted to get ice cream after our tacos. I was already full and didn't really want ice cream. But I got it anyway. I ate some, even when I wasn't hungry. Then I did something I didn't know was possible: I threw it away before I finished. I didn't want it anymore. No stress. No overthinking. No guilt. Just . . . neutral. Who have I become???!!!"

Another student, Lupita, told me that she recently had an experience where she was out to dinner at a restaurant in her neighborhood that she frequented and felt very happily full from her meal. She had stopped eating naturally, which was already a big change for her, but was definitely full and didn't want any more food. Then the restaurant brought out a free dessert to thank her for being a regular. "Before, I would have freaked out. I either would have not eaten it and felt so rude, or eaten it and felt horrible about myself. But this time, my husband and I, both full, ate as much of it as we could. I was so stuffed, but for the first time I understood that being stuffed is okay. I didn't feel scared or guilty, and it didn't drive me to binge later. This really is an amazing change."

EAT WHEN YOU'RE NOT HUNGRY EXERCISE

This week, purposely eat when you are not hungry at least once a day. I don't care how much it is. I don't care where or when or why or how. Just eat when you're not hungry, and see what it feels like.

Eat when you're not hungry and pat yourself on the back for neutralizing food. It's just food.

I know this sounds like a deeply irresponsible exercise, but if you actually do it you'll realize that eating when you're not hungry is just . . . not a big deal. When you're allowed to eat when you're not hungry, it's not even freaking fun.

THE TRAP OF MINDFUL EATING

Traditional "mindful eating" is all about *slowwwwly* paying attention when you eat, with awareness, and being in tune with your body while you are eating. The focus is usually on *slowness*, so you are aware of how it feels to eat, be hungry, and be full.

If you think about it, intuitive eating doesn't *have* to be slow. It just has to be instinctual, but it is often taught through *slow, mindful* eating. And because of that, the terms "mindful eating" and "intuitive eating" are often used interchangeably.

The Fuck It Diet is recommending exactly what it implies . . . Fuck it. Really. There is a good reason The Fuck It Diet helped me, when intuitive and mindful eating kept me stuck for six years. And it's because slow, mindful eating is *easily* adopted by disordered eaters as just another way to obsessively try and curtail what they are eating and control the size of their bodies. *If I just eat really slowly and mindfully and intuitively, then I'll eat the perfect way and everything will be wonderful and I'll be beautiful and loved and approved of forever and maybe even marry a prince and be as skinny as Kate Middleton, who knows. Who. Knows.*

Don't worry about eating slowly or mindfully or intuitively. I have seen, time and time again, that *Fuck it* is the easiest and truest way to get to real intuition with food. There is nothing inherently wrong with eating slowly (or Kate Middleton) or paying attention to how your body feels when you eat. Mindful eating and intuitive eating are great in theory. *But* when you give people who still want to be thin more than anything else the task of mindful eating? Ooof. It just doesn't work. It just adds another rule and backfires big-time. It just becomes another sneaky diet to eventually rebel against.

Sometimes mindful eating is taught by "rating" your hunger on a hunger scale of 1 to 10, but I purposely do not teach it or recommend it. You do not need a hunger scale to learn whether you are hungry or not. The Fuck It Diet's goal is to get you to eat *normally*, by following what you want to eat. You're allowed to follow cravings and even impulse. You do not have to overthink it. It's also encouraging allowing yourself to be full, fed, gain some weight, and allow that biological process to happen without micromanaging and overthinking it. Trying to eat slowly and perfectly will backfire: you'll easily start obsessing and spiraling into overthinking, weird habits, and most certainly, bingeing.

I actually thought I was intuitively eating for *years* before I realized I was still obsessed with weight the whole time. What I thought was "intuitive eating" was just another kind of diet. One of the versions of this was thanks to that book *French Women Don't Get Fat*. I was so excited to learn to eat beautifully and chic-ly and become more beautiful and tiny.

I ate everything really, really slowly. Like *strangely* slowly. I ate lots of yogurt—very slowly. I would make myself starve in between meals. I was still afraid of sugar. I drank a *lot* of coffee. I drank a *lot* of wine. And I cried a lot.

I also wore a *lot* of scarves, but that is neither here nor there. The point is, I thought I was eating what I wanted, and doing it all "intuitively," but I was still judging it and controlling how much. *French Women Don't Get Fat* also encourages you to only eat a half of a banana instead of a whole one, because apparently bananas have grown exponentially over the years, and a whole banana is *just too much*. Absurd.

Friends would go out for ice cream and I'd get a small bowl and be petrified of eating more than a few bites. *Wait, I don't know if a French woman would actually eat this whole thing?? I think I'm only like a 5 on the hunger scale . . .* Oh yeah. Very intuitive.

I would go back on stricter diets in between bouts of my pseudo-intuitive eating. You know why? Because I was still sometimes bingeing. I would still "go off the rails." I'd go crazy and eat a whole banana. And then follow it up with a box of cereal and a jar of peanut butter. I would still rebel against my perfect French/intuitive eating *because there was still something to rebel against.* There was still a way to do it wrong; I was still living my life for being thin. Everything I did was in the hopes that I would be thin. *Even when I thought* I was learning to eat intuitively, the goal was still weight loss, therefore the way I was eating was always with that unspoken rule to try and *eat the smallest amount possible* so I could lose weight.

The realization that changed everything for me was that my obsession with thinness was messing up my eating. We aren't supposed to live our lives scared of food and weight. We are fucking adults who need more food than a hard-boiled egg for breakfast and a salad and Diet Coke for lunch.

Enough with slow overthinking. Those rules will control you. You will never be truly mindful or intuitive if you're operating under disordered food rules. You need to go headfirst into eating whatever you want in whatever quantities, not worrying what number of "fullness" you are.

Under normal metabolic conditions, with ample food supply and *no* restriction or weird slow-eating rules, normal eating happens naturally. Once your body and mind have neutralized food, you'll begin to eat intuitively without forcing anything. The body *wants* to eat intuitively—that's the whole point of the word. It becomes second nature.

WHY ARE WE EATING THE SMALLEST AMOUNT POSSIBLE!?

One of the biggest myths, and unspoken diet rules, is that we should always be aiming to eat the smallest amount possible. We have a subconscious belief that eating the smallest amount is the most responsible way to eat. We believe that people who barely feed themselves live forever in health and happiness, and people who eat heartily die ugly, ugly deaths. And it's bullshit.

What we have learned about calories and food amounts is mostly wrong and sensationalized. And if we have learned anything else from the Minnesota starvation experiment, the amount we have decided is acceptable for diets is absurdly low and dangerous.

You can also use this simple barometer: if you are bingeing, you are still not eating or *allowing* enough. I know we all think that bingeing is the big problem, but there's a new wave of treating eating disorders and disordered eating that is beginning to understand that binge eating is not a stand-alone disorder. It is a *reactive* disorder. And it is reactive *to diet culture* and diet mentality. Bingeing is *not* your enemy. Yes, of course we want to make sure that we have ways to cope with our life and our emotions (much more on that later). But hunger, food, and all the binges you have ever been on are *not the enemy*. The binges are often actually *just* getting your calorie amount to where it needs to be on a therapeutic, restorative level.

There was a time when I thought that eating a full 2,000 calories a day was *totally blowing it*. I look back at my food journal from high school where I had decided, thanks to whatever magazine I was reading at the time, that 1,800 calories was the *maximum acceptable amount*, which of course was nearly impossible to stick to without ending up bingeing and feeling horrible about myself.

So what's up with that 2,000-calorie label? As Marion Nestle wrote in the *Atlantic*, "Despite the observable fact that 2,350 calories per day is below the average requirements for either men or women . . . Nutrition educators worried that [listing 2,350] would encourage overconsumption."

Everyone is so afraid of "overconsumption" that they give bad, extremely low-calorie advice—because everyone's become so afraid of human appetites. And of course, at the end of the article, Nestle gives advice on how to eat the "right amount," so yes, I'm also cherry-picking, but for good reason: *most people are a little disordered about food.* Many, *many* people who go into nutrition go into it because of a preexisting obsession that feels legitimized when it becomes their profession.[29]

Most of us take part in the collective belief that we need to be eating less. We have accepted that we should diet or tightly control the portions we eat for the rest of our lives. But I want you to really think about that for a moment. . . . Is that actually true? Because realistically, there is nothing natural or sane about going through our lives trying to eat the smallest amount possible. No matter what celebrity health gurus say, constantly eating the minimum is *not* good for you.

It is totally crazy from a biological standpoint. The generations before us were people who worked so hard for the food they put on the table. They knew how important nourishment was.

What would your distant ancestors think of the way you feed yourself? They wouldn't understand the way we think about food and weight at all. Carrying extra weight is a biological advantage.

Imagine our ancestors being able to witness us sitting in front of our food, praying to the skinny-gods that we won't eat a lot of it, watching us spend the whole meal trying to eat just the perfect amount and not get too full. Watching us eat just enough to not be hungry anymore. Watching us decide we should be finished and get up from the table still a little hungry, wishing we could eat just a little bit more, and then throw away the leftovers. They would think our entire culture was crazy and insane . . . because it is.

Controlling the food you put in your mouth is not how we are wired. It is not how we were made. It is not how we ate for our entire existence through history until very recently. Nobody finished a meal before they were fully fed because of "portion control." Portion control was, historically, the unfortunate side effect of not having enough to eat.

The body is designed to guide us to easily eat the right amount. It is designed to maintain a stable weight, regardless of how we are eating that day or week. It's not supposed to be hard or stressful, *unless* it is a famine, or times of not having enough food on the table . . . or a diet. Then everything becomes about packing away food and packing on pounds.

My student Mary thought she would devour the world if she ever let herself stop controlling her food—which is a super common fear. But three years after eating as much as she wants on The Fuck It Diet, she really couldn't care less about food anymore. "I just don't think about it. I forget a lot, and I feed myself by second nature. It's just not a big deal. Eating and meals happen, and I barely think twice about it. This is a huge change in my body's and brain's relationship to food. I used to think I should go into food writing or go to culinary school because all I wanted to think about was food and what I was going to eat. This dramatic change only happened by deliberately feeding myself everything I wanted, for a long time."

I am going to keep reminding you that if you have been restricting *at all*, your body is compromised and needs lots of food to heal it and get it back to normal.

I do not recommend that you count calories, because that's the old paradigm that we want to dismantle. But what I do recommend is completely switching around the way you look at calories. Start making sure you are eating *more than enough* to heal your metabolism. Start seeing *more* as better. Start understanding that the more you eat, the more you will be nourished and tided over until the next meal, the more grounded and focused you'll be able to be, and the more calm your body and blood sugar will become, ongoing.

Feeding yourself a lot is going to teach your body (and your mind) that there is enough food, the famine is over, and the fixation on food can end.

FAMILY AND FOOD

There are so many different ways we learn to relate to food. And a big one of those is the way our family relates to food. Surprise, surprise! Some people's families are all on diets, afraid of food and gaining weight. Some people's families are all about cooking and feeding for each other, and that's how they show love. Some people's families are genetically fatter, or have lots of shame and fear around eating. Take time to answer the following questions:

How did your family relate to food? Grandparents? Parents? Siblings? Extended family?

What are the comments you heard family members saying to other people, about other people, or to you, about food or weight?

Try to remember a particularly loaded or emotional situation with food that happened in your family. Describe it and try to remember what that felt like. Do you have any beliefs lingering that have to do with that experience?

What are the aspects of the way your family relates to food that you'd like to untangle from?

What are the aspects you'd like to keep?

THERE IS NO PERFECT STOPPING POINT

Before The Fuck It Diet, I was still caught in the trap of pseudo-intuitive eating. I used to get truly stressed over figuring out when

I was full (thanks, hunger scale!), and I was very concerned that I was going to get *used to* eating too much.

This was the result of the obsessive, fearful side of intuitive eating. I thought that the most important thing I could do, as an intuitive eater, was to pay close attention to my hunger, rate my hunger every few bites, and stay constantly vigilant, bite by bite, just in case I became full mid-bite . . . Again, the unspoken rule that we must eat the smallest amount possible was ruling me from the shadows.

While it is perfectly fine to pay attention to how you feel while you eat, the way that I was focused on stopping points was soooo obsessive, fearful, and unhelpful. I believed that there was *one perfect stopping point* that I should always be finding during a meal, one perfect point of satiation—not too full, not too hungry.

This perfect stopping point does not exist. It's just another eating myth. Instead, there are *many bites on a scale* of comfortable stopping points. And any point is okay. Not only that, but your body can handle a few *extra* bites, or more. It can handle extra food. You may be slightly fuller, but as long as you are feeling and enjoying, you are doing great. Your body always makes up for more food with slightly longer satiation, slightly less hunger for your next meal or snack, a faster metabolism, or some combination of these.

There is no virtue in stopping just before you get full or expecting there to be a "perfect point" every time you eat. That is just going to leave you frustrated and obsessed. Just enjoy your food, goddammit.

THE TRAP OF SEEKING BALANCE

The body is always seeking balance, and after years of dieting, balance is eating whatever you want and need. After restriction, balance is eating a lot.

Balance is completely relative. Balance is the swinging of the pendulum. The further out of balance you were, the further in the other direction you will need to swing in order to *eventually* land at what we think of as "balance." Expecting this process to look like stereotypical middle-ground balance isn't actually balance at all. The idea that *forced* "balance" is balance is bullshit.

Months and years down the line, depending on your own particular journey, body, needs, and timeline, balance will begin to look different. Maybe it looks like salads, sandwiches, and dessert. Brownies and kale. Tuna and mangoes. Smoothies and cake. Whatever. I don't care what you eat. I don't ever care what you eat as long as it's what your body is telling you it needs and you're listening. I just want you to be happy and intuitive and to chill out about it.

One reader said that it took a really long time to stop eating more than they physically wanted at night. "I'd eat so much more than I even wanted to at night. It really stressed me out. Then one day I realized that my whole dieting 'trick,' for *years*, had been trying to go to bed hungry. I'd force myself to go to bed hungry. I finally realized that what my body was doing with this eating at night was trying to overcorrect for years and years of hunger at night. I was shocked by how smart my body was. And once I truly accepted this and surrendered to it, things started to shift, and my hunger at night became calmer."

Forcing yourself to "just eat balanced" or "just eat everything in moderation" is NOT balance after you've been dieting for years, so just put that out of your mind, and trust your body. Your body is seeking balance even if eating a whole box of cereal in one day isn't how you imagined balance to look.

OUR GOAL IS TO NEUTRALIZE FOOD

I keep talking about *neutralizing food.* So let me talk about what that actually means, and what that may look like for you. When food is neutral, it has no morality, no judgment, fear, or guilt attached. It is just food. When food is neutral and free of judgment, it becomes so much easier to listen to your cravings and begin to eat intuitively.

There may have been a time in your life when food was neutral—not good or bad or stressful—and when you were hungry or craved something specific, you ate it. It was no big deal. This may have been when you were a child, or maybe you were even food-neutral early into adulthood, but *that* is the relationship with food we are trying to get back to: food ease. That is the only way you can eat normally and attempt to listen to your body in any sustainable way.

But even if you've never experienced neutrality with food, I still promise it is possible to get there, because *I* had never experienced it either.

If I ever was neutral about formula or baby food, I certainly can't remember, because all I *do* remember is that my number-one goal in life was to *trick adults into giving me snacks, because god knows there are only carrots and fiber cookies for snacks at home.*

I don't remember a time when I was neutral about food. I went straight from childhood binge eating and snack obsession and trickery to teenage dieting and binge eating, which stretched into my mid-twenties. My strongest memory of vacations were the pancake breakfasts at hotels, my favorite thing to do at friends' houses was to eat their snacks and fruit gushers, and my memories of holidays were the cookies that distant relatives would give to me and my brother. My mom would scoff because *her children were so obsessed with food! Don't give them even more junk than they've already managed to stuff themselves with!!!* We definitely *were* obsessed with food, so I didn't think food neutrality was even a possibility for me, because it seems like I was *born* a food addict.

But this is what the brain does under real (or perceived) restriction. The hunger hormone rises, and the brain fixates and obsesses on food. My mom was just trying to make sure we were healthy, and seeing her kids *obsessed* with bingeing on candy, she figured she had to double down, which is pretty normal parenting. But I argue that our obsession was perpetuated by feelings of restriction, and got us into a vicious cycle, and made us more and more insane around food.

It turns out that the only way to neutralize food is to allow it all, always, forever. When I decided I was going to give up restriction and "eat my way to the other side," I was still petrified of gluten and industrial seed oils. So, I decided I would "eat myself to the other side" first with potatoes and butter. Just adding good quantities of carbs alone was enough for me to tackle emotionally.

I added in lots of different carbs and trusted what I was reading about healing from diets: my body and metabolism needed these carbs. I was very, very hungry, for months and months. I ate a lot. I had been bingeing right up until I started The Fuck It Diet, so part of my brain would try to argue that I had no right to be this hungry, but the other part knew that this was the way it had to be.

Slowly but surely, the more I would eat a certain food, the less I would crave it and care about it. The more I ate it, the less I would go nuts over it. Eating it freely *literally* lessened the power, and I started easily stopping when I wasn't that hungry anymore.

I ate potatoes, bowls and bowls of granola and cream (yes, cream), lots of fruit, milk, and ice cream (I had been diagnosed lactose-intolerant at age two, so these were big leaps). I would eat my nighttime food slower—not to eat less, but to prove to myself that I was allowed to be eating. I would make it a long, slow luxurious feast to prove to myself, *This is not a binge. This is fully allowed. All of it.* Sometimes I would end up eating less than a usual binge, sometimes I'd eat basically just as much—I just was so much calmer and knew it was all good. It was all allowed.

Not too long after, all those foods—milk, ice cream, granola, potatoes, all the fruit I was afraid of from being paleo, and whatever else I ate—became really neutral. They just didn't have a pull over me anymore. I was allowed them. I had been eating them. I had been enjoying them. And I never had to stop. It wasn't actually until five months in that I decided to eat a piece of bread. Then I started eating nachos whenever I could at restaurants. The oils they fry them in had really freaked me out, so eating them was my version of exposure therapy. Also I just really love nachos.

As I allowed myself to eat more things, and more and more food became neutral, my body started being able to actually pick and choose and crave what it actually needed. Sometimes milk and potatoes, sometimes bread and cheese, sometimes fish, sometimes a Chipotle burrito, sometimes fruit, sometimes . . . *whatever.* I don't remember all the foods that exist right now, but I ate them.

You can't trick yourself out of this one—you actually have to allow yourself the foods that you crave. Your body and brain are not idiots. They'll know if you're lying. You have to go through the process. You can't *pretend* to allow foods. You can't pretend you're allowed to eat dessert but secretly have rules for how many times a week is acceptable. Feeding cravings allows the craving to pass. Your body and mind crave things that are forbidden or restricted, as well as things they need, so once they're not forbidden, cravings become more and more in tune with what the body really needs. At this point, more than seven years into this journey, I really just want to *feel good* from eating, during and after.

The food-neutrality process happens at different rates for everyone. I recommend, if you can, to do this as quickly as possible. Let go of control of as many foods as quickly as you'll let yourself. You will eat yourself to the other side faster the more you fully allow it.

But if you are like I was and find yourself still afraid of arbitrary ingredients, just do it in chunks. But you have to start.

There is absolutely no substitution for *actually* eating the foods that scare you.

Remember, the goal is not to eat food in order to get to a place where you don't want food anymore. That's never going to happen. But you *can* take the charge and power out of foods by allowing them. The goal is to get to a place where you only think about food when you need to: when you're hungry, grocery shopping, planning a party, cooking, etc. If you're cooking for a family, navigating your cravings with your kids and partner will probably always be a juggle, but maybe now that you can eat macaroni and cheese, some nights will get easier.

WILL I BE EATING THIS MUCH FOREVER?

Most people who have controlled their food in the past have a little freak-out when they see how much hunger they have. You will probably need a *lot* of food, and you will most certainly freak out about it. You may assume that you are the one person doing this wrong. You will think that you are the one person who is broken, beyond saving, the one true food addict who will never, ever stop obsessing over food.

You will think, *Not dieting anymore is one thing, but the amount I'm eating can't possibly be normal or good!* And I *wish*, for your sake, you could always remember the biology of a famine, and how normal it is to heal your body with large amounts of food after a long famine.

You will not be eating like this forever. How long were you on a diet? Two years? Five, ten, thirty years? If you expect to be able to reverse what dieting has done to you in a few days, or even a few weeks, you're not looking at the big picture, and you will just be setting yourself up to feel disappointed and scared.

The amount of time it takes to stabilize both eating and weight is different for each person. It is different depending on how long you've been dieting, how much you trust and allow this process, what your weight set range is, how close you currently are to that range, and how temporarily compromised your metabolism is. It just depends. The average is a few solid months to get over the physical hump, then many more months to reverse the mental and emotional parts. But there is nothing I can write or teach that will tell you exactly what will happen and when.

However, I know it can be helpful to hear other people's experiences. My student Sarah shared,

> I've been on TFID for about 3½ months now and have eaten and eaten and eaten and it's been great. For the first time I've found myself thinking I might actually be a bit "fooded out." All I can think tonight is *Nooooooo, no more!* It feels a bit weird! I really wouldn't have believed that was even possible six months ago.

My other student Nicole said,

> I went through a long phase of eating just *everything* in huge quantities. And then I got to a point where I was just like . . . *huh . . . I just don't WANT that peanut butter cup* and

it's *so* different from all those years of restriction and "bad" foods. It just *happens*. It feels easy, like a genuine intuition I never experienced before.

Allie just reached out and told me,

I started this 3 months ago. At first I was eating a lot of drive-through burgers with bacon, large dishes of pasta, and Dutch pancakes covered in maple syrup—all of the previous no-no bad-foods. . . .

A few weeks ago, things started to change. I started to crave things like blueberries, potatoes with skin, steak, and . . . kale. And at around the same time, I started to get the feeling from my body that I did not want to be overfull anymore, and for the first time it wasn't this shameful reaction or a punishment or repentance for overeating. . . . It was just like *Okay, I don't need to do that anymore*, and I naturally stopped.

I can't believe how easy and intuitive the change is, and it really only happened because I truly surrendered to all food, in all amounts.

Most people want this phase to go quickly, because they're worried about weight gain. That's what gets us into this mess in the first place—the fear of weight gain—and it is also the thing that can keep us stuck. All I know is that the *faster* you surrender to this, the faster it will all happen. If you dip your toe in and out for a year, the water will keep feeling cold. But if you jump in and trust, you'll float and you'll get used to the temperature faster. Is that an annoying metaphor? Oh well.

If you want to get to a place where food doesn't rule your life, you have to surrender to food. It's scary. I know. I understand that you just want it to be over and done with. But see if you can just *enjoy* this time of needing lots of food and letting yourself eat what used to be forbidden. It's pretty fun if you can remember to not be terrified of every bite.

I wish you could always remember the Minnesota starvation experiment, and how many thousands of calories they put down every day in the many months following their 1,600 calorie "diet."

I wish you could always remember how *normal* the large amounts of food are, biologically, and remember how much your ancestors would eat in times of plenty, and remember that eating a lot is healing, nourishing, and the only way to get to that place where your body and mind don't feel deprived.

I wish you could realize that even if you think at times you are "irrationally hungry," that listening to your body is *never wrong*. I wish you could always trust that. But you won't. You'll freak out. And hopefully you will turn to this page of the book and read it again, calm down, smile, keep eating, and then go to bed.

NO FOOD IS OFF LIMITS

Another essential part of this process is allowing yourself to *eat all foods*, from now until the end of time. That includes foods that scare you, foods you have declared "off limits," foods you believe are very unhealthy or "empty calories," foods you are afraid will make you gain weight, and even foods that are objectively "shitty" and "fake."

The reason you must allow food "now until the end of time" is that if you tell yourself something like, "Oh, well, I will let myself eat whatever I want for a month and then it'll heal my eating and then I will 'eat healthy' after that," that is actually a form of restriction. I call it mental restriction, and it'll mess up your eating just as much as physical restriction.

That's also what I call an *impending diet.* And impending diets, and any other *conditions* on your eating, are the opposite of The Fuck It Diet. We will go really in depth about mental restriction later on, but I am introducing it now to explain why it is so essential to let yourself eat every single food, no conditions, no impending diet, forever.

We also have a tendency to want the foods we are not allowed to have. Don't give yourself the opportunity to crave something *just because you don't think you're not allowed to eat it.* Lots of people gravitate toward childhood comfort foods at the beginning of The Fuck It Diet. Let it happen. Enjoy it. When you are actually allowed to eat any food, those foods lose their power and you tend to crave and eat them even less, especially when you don't need or want them. Then you'll only eat them when you *truly* want them and with way less drama. This is the great paradox of allowing all foods. Your subconscious internal rebellion is neutralized.

Most people feel like if they let themselves start eating a food they have no control over, they will just never stop eating that forbidden food. And hey, as long as that food is forbidden, you are probably right. But once you allow that food fully, you'll see the power fade. Yes, you may eat a lot of it, but all for a good cause. The only exception to this rule is with foods you are truly allergic to, like a peanut allergy or celiac disease. In that case, not eating foods that give you an immediate reaction is self-care. You can and will learn to tell the difference.

It's also important to point out that certain foods only have the power to "make you gain weight" if you are deficient in one of the macronutrients in that food—like fat or carbs—and the body feels malnourished. In that case . . . yeah, you will hold on to weight once you finally feed yourself, as you are supposed to.

You may have noticed that when you've been on a diet, and gone "off the diet," you gained weight quickly. That is normal, and it happened because the body was so relieved that it finally got you to eat what it was lacking. Your body immediately packs on weight after a diet in order to repair itself and store fuel just in case the famine continues. It's simple survival.

But of course we assume that *we* are just *the kind of people who gain weight quickly*. But it is all *because* of diets. It's *because* of us trying to micromanage our intake of certain foods or macronutrients (fat, carbs, and protein). There is no way around this part of the journey. The only way out of this cycle is to allow all foods once and for all, and to actually eat them—especially ones you are afraid of and have avoided for a long time. It is only feeding yourself that can stop the cycle. It's only feeding that will stop the body from freaking-the-fuck-out, and let you stabilize at a healthy weight for you without yo-yoing every time you eat a piece of toast.

WRITE A LETTER FROM YOUR BODY

This is a great thing to do *always* and anytime. Whenever you're in doubt, whenever you're panicking, whenever you need some direction, or wisdom, write a letter to yourself from your body.

Yes, you need to use your imagination to do this one. But just assume that you can tap into how your body feels. What does your body think? What does it want? What is it thankful for? What is it frustrated by? What does it need? Write for five to ten minutes from your body to you. And you will be amazed by the insight you get. This is a quick snapshot into your intuition.

YOUR CRAVINGS ARE YOUR FRIEND

It has been drilled into our heads that cravings are bad, and that we should try to eliminate them because they are sabotaging us. So we try to become cravingless machines, constantly exercising and shopping for smaller and smaller clothes, eating lean protein and green beans until our delayed but inevitable deaths at ninety-eight. Sure, we eventually die, but at least we are skinny! At least we stuck to our diets and didn't eat that pudding they tried to force on us at the nursing home.

The truth is, your cravings are incredibly important. The

kinds of foods you deny yourself are the kinds of foods your body ends up *needing*, and therefore, craving. Yes, even desserts. Carbohydrate and fat-rich foods are often what your body craves because *calories are what will heal your metabolism the fastest.* The dense and easily available energy is what will save you from a famine state the fastest. That means that all those years craving desserts while you were on diets wasn't your body failing you, it was actually your body trying to lead you in the right direction.

Remember, if we weren't in our own way and didn't think we needed to diet, we would follow cravings and hunger, heal our metabolisms, and get out of famine mode quickly—and eating would normalize within a few months. But since we resist it, the body speaks louder and louder, until we eat the entire pantry shelf of desserts and decide we must have a sugar addiction.

Ray Peat, who has a PhD in biology and specializes in physiology, has done extensive research showing how helpful sugar is to the metabolism and stress response. He says, "Any craving is a good starting point, because we have several biological mechanisms for correcting specific nutritional deficiencies. When something is interfering with your ability to use sugar, you crave it because **if you don't eat it you will waste protein to make it**." You know those stupid charts that say, "This is what you're *really* craving," and next to chocolate it says "You don't need chocolate, you just need magnesium, so eat 12 almonds instead!" Fuck no. Fuck that. When you crave chocolate, even if you also need magnesium, you probably need carbs too, or *else you would just crave almonds.* Eat the chocolate.

People have some pretty strong health beliefs and fears around sugar that can keep them from giving in to cravings. When I started working with Sam, she *really* wasn't on board with giving in to her cravings, because for years she had been convinced that her sugar cravings were because of an overgrowth of candida. Sam was sure that giving in to her cravings would only make her cravings for sugar *worse*.

Candida is a yeast that is a normal part of our gut bacteria, but many people fear that eating sugar creates an overgrowth of candida. Good news: it doesn't.[30]

But I spent years petrified of sugar and candida myself. I'd eat versions of a "no-carb" or "low-carb" diet to try and "kill candida." And of course, as dire as it all felt, it didn't seem to work or help, and I couldn't keep it up for very long anyway. I felt doomed. I felt like I had been invaded by candida, but that I didn't have the kind of willpower to keep up the cure. It was really disheartening and exhausting.

But this is a great example of how we are trying to heal ourselves in too focused of a way, ignoring the bigger picture and our whole body and being. First of all, we always have candida and it is a necessary part of our gut flora. But our gut flora can become imbalanced and exacerbated by deeper systemic imbalances. For instance, many people with yeast overgrowth actually have an underlying heavy metal toxicity,[31] where the yeast is actually helping to absorb the metal and *protect* them from more acute heavy metal poisoning. So, if you cut out sugar, the underlying cause (heavy metals) is still going to be there. You're not curing anything long-term. You're just giving yourself a miserable year of eating almonds when you crave chocolate.

(And FYI! Another example of a systemic imbalance is a slow, impaired metabolism—one that has spent years in a diet survival state.)

You also can't "starve" candida by eliminating sugar, because if it's not getting sugar in the gut, it can move to your blood, where there is always sugar to keep you alive (if you aren't eating sugar, your muscles are being broken down to keep sugar in your blood and cells). Plus, for example, honey has actually been shown to help curb candida overgrowth because of its antifungal properties. Not honey extract, or a drug derived from honey . . . *honey. Sugar.* My point is that we are chasing the wrong demon. Just because candida (like every cell) lives on sugar, we assume the cure is starving *every* cell in our bodies? Your best bet is feeding yourself and supporting your body and your metabolism with food and carbs and probiotics.

So, remember Sam, who was sure she was addicted to sugar because of candida? She finally let herself eat carbohydrates, and to her surprise, her carbohydrate cravings eventually normalized. She was able to tell when she had had enough. And she had been *convinced* this could never happen because of her *candiiidaa.* I have heard so many stories from students and readers saying similar things—that they were die-hard sugar addicts who believed sugar was a drug that was ruling and ruining their lives, only to go on The Fuck It Diet, allow and eat sugar, and voilà . . . no more addiction.

Jackie said, "I used to DEEPLY believe I had a sugar addiction. And I was in and out of Overeaters Anonymous for years. But once I stopped restricting I realized that all along it had just been the restriction talking. I still enjoy sweets now, but not like I used to AT ALL. I no longer feel like certain foods have power over me."

You are still seeking health on The Fuck It Diet. I just believe that the cause and the cure lie in a way more holistic approach than we usually take. We've believed for so long that indulging our cravings is irresponsible, even though it is actually one of the most essential steps toward truly listening to our bodies. Renourishing and mineralizing your body is so important after years of whacked-out eating. Healing your appetite and metabolism and finding true food intuition is a *huge* step toward sustainable health.

Eating a varied diet, with lots of fruits and vegetables and nourishing meats and fats and carbs, full of vitamins and natural fiber, is all great. Learning about sustainable farming, organic food, and humane meat is all great stuff too. Food and herbs are magical, healing, nourishing, and exciting. I'm all about it. If learning about all of that feels good, go for it.

But healing your metabolism and relationship to food and weight should be the *first steps*. They are the most imperative steps if you want to be able to fully enjoy eating and the body you are living in.

EAT THE FOODS YOU HAVE NO CONTROL OVER

If you are worried you will never stop eating a certain food you always crave, the quickest way to see that shift is to allow that food completely—in whatever quantities you crave.

What foods do you feel you have no control over? Popcorn? Brownies? Tortilla chips? Ready for this?

. . . I recommend that you choose one, and then eat that food in the high quantities you crave. Surrender to eating the food you are afraid of. Hey, you can eat it for breakfast, lunch, and dinner if you want, and see if your relationship to that food changes. I bet you it will.

IN DEFENSE OF CARBS AND SUGAR

As I've already alluded to . . . I used to believe that carbs were Enemy #1, and that they would destroy my health and hormones, and turn me into even more of a pimply disease bag than I already was. I went on low-carb diet after low-carb diet. (Except when I was raw vegan. Then I would eat seven boxes of dates at once, but that's a whole other rabbit hole . . .)

Our fear of sugar is *deeply* entrenched in our culture and our psyches. Our current beliefs about sugar are based on studies that, again, aren't always looking at the whole picture but have been turned into fearmongering headlines. For instance, the assumption that sugar causes diabetes and insulin resistance is a classic case of *correlation is not causation*.[32] Impaired sugar metabolism is the *result* of diabetes, not the cause, and once a patient has diabetes, limiting sugar *may* help mitigate symptoms, but it is not the cure. In fact, not eating enough carbohydrates can even make diabetes symptoms *worse*.[33]

Then there's the idea that sugar is more addictive than cocaine. That one made for a very catchy headline. The truth is more like this: there was a college study where rats chose Oreos over cocaine. (There was also a study where rats would essentially commit suicide with cocaine *if they had nothing else to do*. But if they had other things to do, and other rats to party with, they would ignore the cocaine and live their damn lives.[34]) But what I'm really trying to say is . . . rats eating Oreos over cocaine is ignoring way too many other factors, and just salacious clickbait.[35] And thankfully other studies agree.[36]

Next: Sugar feeds cancer cells? Well, sugar feeds *all* cells, including your entire body and your brain. Sugar and ADHD? No.[37] In fact some studies found a slight improvement in attention.[38]

But these are all new things I have learned, *since* The Fuck It Diet. Back in my diet heyday, I was convinced sugar was the worst thing I could eat, and I felt like the proof was in my *addiction* to it. Carbs were a gateway drug to *more* carbs and *more* carb cravings.

And even during those times I thought I was "following my body" and eating intuitively, I was *still* monitoring carbs and sugar. Always, always, always. I never, ever, made myself grains or potatoes. I would try to make everything vegetables and meat. And I would order the least carby meal on the menu if I ate out. I was a *master* at that. I would eat 90-percent chocolate and convince myself I loved it. I would eat almond butter as dessert, and never feel satisfied, so I'd end up eating half the jar that night.

No more than two slices of bread. Starch or dessert. Rice is a waste of calories. And on and on.

Even when I wasn't on a formal diet, I was still operating under old diet rules. These rules were quiet, and almost undetectable because I had accepted them as truth. But they were completely in control of me.

And guess what? I was *always hungry*. I was so concerned with trying to eat the smallest amount possible, *especially* the smallest amount possible of carbs, that I got hungry almost as soon as I had decided I should be finished.

I *believed* what I had been taught about carbs. I thought that my body would get unhealthier the more carbs I ate. Doctors casually told me that too, so I thought I was being responsible. No matter what else I tried, I would always come back to low-carb diets, convinced it was the one thing that would heal me.

I also used to think sugar was the reason I was addicted to food, and that avoiding it would take away my cravings. *Sugar was the cause of every problem. Sugar was the reason my hormones were out of whack. Sugar was the reason I wasn't skinny.*

But the more I cut out sugar, the more it all became a self-fulfilling prophecy. The less I ate sugar, the more I craved it and the harder time my body had processing it when I would inevitably eat some. My body's impaired processing of carbs and sugar seemed to prove those beliefs about sugar even more. And do you know why????? Because feeding yourself the smallest amount possible, and *willing* yourself to eat a very small amount of carbs at every sitting, makes no biological sense.

We think that restricting carbs and sugar will help us burn

fat and become fit and healthy, but what is actually happening is a crisis mode run by stress hormones that creates inflammation and a slow metabolism. You will burn fat in the beginning, but *not* sustainably.

And now I will go into *the most science* I will ever go into in this book:

Whenever our output exceeds our input—meaning you aren't eating enough, resting enough, or eating enough carbs—the body releases adrenaline and cortisol, which are two major stress hormones that help the body create fast fuel for your cells. Without this fuel, we die. Sugar (glucose) is the most efficient fuel source for our cells, because it uses the least oxygen, makes the most usable energy, and creates the most carbon dioxide—which removes calcium and sodium from cells, keeping them stable.

The first stress hormone, adrenaline, finds glycogen in the muscles and liver to burn for fuel. After that, adrenaline will burn fat, which, by the by, is *not* good for your health or metabolism, because adrenaline uses three times as much oxygen to burn fat for fuel, creating less carbon dioxide and less energy—and also creating inflammation.

The second stress hormone, cortisol, will pull amino acids from the skin, thymus, and muscle, which are taken to the liver to be used for energy. This lowers thyroid function, digestive juices, body temperature, and pulse. Simply put, restriction of calories, or restriction of carbs, lowers your metabolism and creates inflammation. No bueno.

Sugar is an unfairly demonized macronutrient. Sugar is really just pure fuel that keeps us alive minute to minute, and it is required in the blood at all times (blood sugar). Denying the body glucose requires the body to *create glucose* in a complicated process that raises stress hormones, creates inflammation, and impairs the metabolism.

The less you eat carbs, or try to replace carbs with zero-calorie fake-sugar products, the more likely you will become chronically hypoglycemic. Your body perceives low blood sugar as a stressor, which kicks adrenals into overdrive and pumps out stress hormones. It also assumes you're in starvation mode, which, remember, will make it counteract with ghrelin, the hunger hormone (even *more* on him later!). Meaning, the *less* you eat carbs, the slower your body will burn fuel and the slower your metabolism becomes. And I promise you, that's not what you're looking for.

In this state, you are more likely to then binge and yo-yo, and stay perpetually in a whacked-out glycemic roller coaster for all of your days. You can also think about it this way: sugar is *meant* to be addictive when we aren't eating enough of it—because we need it. Just like oxygen and sleep. And unlike drugs, when you actually consistently let yourself eat it, it has a restorative and calming effect on the body and metabolism because it lowers your ghrelin, which lowers your fixation on food so your body doesn't feel "addicted" to it anymore. Eating carbs allows the appetite to calm down, and is the only way to experience easy, normal eating.

Once I started letting myself eat bowls of rice or pasta or quinoa with no thought to the old diet carb rules, something magical happened . . . I actually started feeling full for more than twenty-five minutes.

I started eating sugar again, full force. Fruit. Ice cream.

Honey. Candy. I even ate sugar *without* pairing it with protein. Imagine that. And now my relationship to sugar is awesome. I did not spiral into a four-year sugar binge like I always feared I would. It was more like a three-month sugar bender, and then it all normalized. Now sugar is just sugar. Eating sugar actually ended my dysfunctional relationship to it. I fed my body sugar, and finally, my body started telling me what it needed.

I eat plenty of carbs every day, but I can tell when my body has enough. I can also tell, mid-dessert, if I'm ready to stop. Not in a creepy, "hunger rating scale" way. Just like . . . *I'm done.* It stops tasting as delicious and I just stop *wanting* to eat. And because I am allowed to eat as much dessert as I want, for the rest of my life, stopping is no big deal.

This is the part that people rarely let themselves get to—the neutralization, biologically and emotionally. Instead, people panic or fight themselves, and stay in the binge/repent cycle. Guilty over eating, trying to cut back, then rebelling against the cutting back—and the cycle continues.

I'm not saying that eating candy and only candy for the rest of your life is a good idea. *You know that.* But what we are told about sugar and carbs is destructive to our relationship to food and feeding ourselves. We need way *more* than just sugar (fat, protein, minerals, vitamins, sunshine, sleep, connection, people, oxygen), but we need carbs and sugar, too. And the less you eat them, the more likely it is that you'll crave them, and the harder your body will work to break down protein and muscle into sugar for your brain to use *to keep you alive.*

Sugar is only addictive to people who are either physically or mentally (and often both) denied carbs or calories. It is a fast fuel that we are wired to crave when food or fuel doesn't seem abundant. Again, sugar is addictive like oxygen is addictive. Your body fucking needs it.

Oh, and 90-percent chocolate is legitimately disgusting.

IN DEFENSE OF DELICIOUS AND DECADENT FOOD

Delicious food is healthy. Please note I did *not* say healthy food is delicious. I said: *DELICIOUS food is healthy.* Butter, salt, cheese, meat, carbs, fats . . . fatty carbs, fruit, butter-drenched veggies, stews and soups, sourdough bread, wine, honey, full-fat dairy . . . all that delicious food is *so incredibly good for you.*

Some of you may have been on the paleo or Whole30 train already and understand that calories aren't "the problem." But I want to talk about that a little bit more. Because even when I was paleo, I still thought that the *goal* of paleo was to *eventually* rely so little on food that I ate less and became a sexy, carnivorous, emaciated meat fairy.

As it turns out, whole, fatty, carby, calorie-filled foods are the kinds of foods our bodies need to eat. They are filled with macronutrients, minerals, and vitamins that we need. Our old belief that calories and carbs are the problem is completely backward. Eating diet food is the thing that malnourished us and deprived us of vitamins, minerals, and the macronutrients we needed. It's the thing that perpetuates the cycle of food fixation. We need calories, and we need carbs, fat, and protein, every day, for the rest of our lives. That 250-calorie meal is a joke. You'd need twelve Lean Cuisine meals a day. Or more.

For anyone afraid of fats, or especially saturated fats, you're afraid of those for the same reason you are afraid of *being* fat: misinformation and scapegoating.

For instance, the war on butter is just another misguided (and potentially sinister) misdirection, and the correlation with heart disease is again . . . just plain untrue. Heart disease was really rare in America in the early 1900s, when people ate lots of butter and saturated fats. Between 1920 and 1960, heart disease rose to become America's number-one killer, during the same period butter consumption dropped from eighteen to four pounds per person per year and we started eating margarine.

You can eat whatever you want—this is The Fuck It Diet, after all—but it is my duty to tell you that margarine was created as a diet food cheaply in a lab, and marketed as a health choice while butter was villainized, and people are *still* clinging to a fear of saturated fat from this misinformation to this day.

Butter contains lots of stuff that protect us from disease. It's the best source of vitamin A, which is needed for healthy thyroid and adrenals; it contains vitamin E, lecithin, and selenium. It's good for your immunity, arthritis, and your gut, and maybe even cancer. Its short- and medium-chain fatty acids protect against pathogens and have strong antifungal and antitumor effects. I actually am having a really hard time choosing *which* of the millions of good things to say about butter, but I truly believe in butter, especially humane, pasture-raised-cow, yellow, yellow butter.[39]

Fat is so important for our hormone function, and these long-vilified animal and dairy fats contain butyric acid, which is protective and restorative to our guts, and helps with metabolic syndrome, insulin resistance, and inflammation.[40] And by this logic, cheese is a health food. You are welcome. You are so, so welcome.

If you're a vegetarian or vegan, you obviously don't have to eat butter or animal fat just because I say it's healthy (and I was vegetarian and vegan for a time, too). Obviously, The Fuck It Diet is not in favor of you forcing yourself to eat anything *because* it's healthy. *But* I *do* encourage dieters and disordered eaters who are vegetarian and vegan to get really honest with themselves on *whyyyyyyyy* they eat the way they do.

At the end of the day, that's all we can ask ourselves, across the board—*why do we do the things we do*? And at this point I have given you scientific reasons why you need ample carbs, sugar, and fat. So *why are you still afraid of cake*?

THROW OUT YOUR PROTEIN BARS

Well, you can keep your protein bars if you genuinely like eating them, but this step is all about going through your kitchen, pantry, and fridge, and getting rid of the food that you only have because you think you "should"–but that you don't actually enjoy eating. The purpose of this step is both practical and symbolic.

If tossing your gross "health" or diet foods feels like too much of a waste to you, you can alternately just take note of the things you bought because you were told they were healthy, but don't enjoy eating, and then never buy them again.

But I vote that you get rid of them. Either throw them out, or donate them to the homeless. But I can't tell if that is kind or cruel.

IN DEFENSE OF SALT

When I started dieting in high school, I heard that I could suppress my appetite by drinking lots of water. So I fucking *drank lots of water because I was amazing at following diet advice.* I pounded water. I could pound a bottle of water in one swoop. I was *amazing at drinking water.* And I started getting thirstier and thirstier. I drank more water than anyone I knew, and I was significantly *thirstier* than anyone I knew. And I peed *constantly.* I thought, "Yes! Look at me! Being responsible with my clear pee!"

I was thrilled that I was so good at drinking water, but it was kind of ruining my life. I couldn't sit through a movie without getting up at *least* once. I was constantly peeing, constantly thirsty, and of course, still constantly hungry. But I was following instructions! I was being a good, water-pounding dieter.

I did this for ten years, and it turns out, this caused overhydration, which leads to a kind of *dehydration*, because it is constantly flushing out all of your electrolytes and minerals, which are essential for nearly every part of your body's function. I was constantly thirsty, but I just kept drinking more plain water, which flushed out electrolytes even more. The cure is not more water, the cure is more minerals—and specifically salt.

This water realization happened at the same time as my Fuck It Diet, so I started drinking water with juice and salt in it to try and rehydrate. It actually turned out to be a profound part of the experience, because it forced me to accept that my pure water obsession was maybe killing me. It went right along with the concept that *we need food, electrolytes, salt, sugar, minerals, vitamins, and nourishment*, not flushing ourselves out with pure H_2O. The answer is *renourishing.*

The good news is that the correlation between salt and hypertension is also based on faulty science. As Morton Satin, PhD, put it, "After World War II, when refrigeration began to displace salt as the main means of food preservation, salt consumption in the U.S. (and somewhat later in other countries) dropped dramatically to about half that rate, or nine grams (1.8 teaspoons) per day and, based on twenty-four-hour urinary sodium data, has remained flat for the past fifty years. During that time, rates of hypertension have increased."[41]

The salt drama is just that: *drama*. We need salt. Salt is a vital nutrient and needed for normal cellular metabolism. It is necessary for the functioning of our nervous system and our digestive juices, for neutralizing food-borne pathogens, for our extracellular fluid, and for our blood and plasma. Not eating enough salt will cause the body to go into a sodium-sparing mode that over time leads to—you guessed it—*inflammation*, including insulin resistance, metabolic disease, cardiovascular disease, and cognition loss.[42] Chris Kresser states, "Animals in a truly sodium-deficient state will seek out salty food and often consume far more sodium than needed to restore homeostasis. These behavioral changes in response to inadequate salt intake further demonstrate the biological importance of dietary salt."

As part of a worldwide study of salt intake and blood pressure called the INTERSALT Study,[43] the Yanomami tribe from the Brazilian rain forest was studied to try and prove that high salt intake causes high blood pressure. The Yanomami tribe have a very *low* salt intake, which the INTERSALT study *correlated* with their low blood pressure and nonexistent cardiovascular disease. There are two big problems with this, though. One, this is just correlation. And two, they also have *dismally* low life expectancies, between twenty-nine and forty-six years. So what are we really going after here with our salt reduction? Chris Kresser said about the INTERSALT study, "When average life expectancy is plotted against the countries' average salt intake, the trend shows that higher salt consumption is actually correlated with longer life expectancy." Boom.[44]

Basically, everything you have ever been told about dieting is wrong. So I'm happy to be the one to tell you that you can and should eat salt, among all the other things. And just like other parts of your appetite, you can trust that when you've had *too much salt*, you will crave water, and vice versa.

Don't freak out about this. Just remember that if you feel like you're peeing a lot, it can be helpful to drink mineral water, or put sea salt, electrolytes, or other trace minerals in your water. This is also a good excuse to act like a kid and drink juice. You're welcome.

IN DEFENSE OF "SHITTY" FOOD

Don't worry, this isn't where I tell you that some foods are better than others. Food neutralization is always our goal. This section is for the people who *just can't stop judging certain foods.* It's for the people who can't shake the scary things they've learned about certain ingredients or additives. It's for the people who are still petrified of the foods they consider to be "shitty." I'm not telling you to start judging food, I'm telling you to lighten up around food you already judge.

Do not obsess over trying to keep all your food "clean" and "healthy." Not to mention that "healthy" means different things to different bodies. So it's *less* important to go into the details of what "healthy" is, and it is *more* important to talk about the disordered, obsessive avoidance and judgment of "shitty" food. And when I say "shitty," what I mean is food that *you* consider to be shitty.

I am all for you eating whatever organic and nourishing food you want and crave and enjoy, but *obsessively* sticking to a whole-foods diet while doing The Fuck It Diet is . . . simply not The Fuck It Diet. You will run into stress and trouble if you try and do it that way. Plus the fact that having the money to afford farm-to-table, artisanal organic foods is just *not possible* for most people—but we can save that for another book and bring this back to disordered eating.

There is a big difference between seeking out food because you genuinely want it and seeking out food because you are obsessed and petrified of the alternative. Fear and *stress* and obsession are always going to get in your way. Obsession with healthy food is unhealthy. Your fear of "shitty" food isn't serving you. It never has.

If you find yourself stuck over your judgment of shitty foods, instead of expecting to change your food fears and beliefs overnight, I think a good, *sane* goal is to *let yourself eat* whatever foods you still consider to be shitty. Even the shittiest of the shitty food you can think of. You can eat shitty food for the rest of your life, and I mean that with my whole heart. Not only are you allowed to eat it, but you're allowed to eat a lot of it, enjoy it, and notice how you *still thrive* even if you eat food that you don't think is healthy. Because hey, even Cheetos will keep you alive in a famine.

My point is: just *eat. Nutrition isn't so black and white.* Your body will eventually become really good at telling you what it can handle and what it can't. My other point by using the term *shitty* is: *even foods you can barely defend are on The Fuck It Diet because . . . chill out.*

Food perfectionism has gotten you nowhere. *Stress* over the food you eat is arguably worse for you than the food itself. Stress has been shown to change gut microbiota,[45] can shut down or slow digestion, and raise inflammation.[46] The lining of the gut is literally a part of our nervous system, and every process in our body is interconnected. Stress physically affects your body, your nervous system, and your bodily functions and processes.

On the other hand, under calmer circumstances, our bodies are wired and equipped to take the good from food and process out the bad. These are all reasons to just surrender during this process and let yourself eat whatever food you want, "whole" or "shitty" or somewhere in between.

My friends from grade school and high school grew up on Fruit Roll-Ups, Lunchables, and Cap'n Crunch, while I was eating organic beans and rice and Tofutti, and almond butter and organic jelly in whole-grain pitas, and guess who ended up with health problems? Me. I don't blame the food I was eating—I'm just saying, health is way more complicated than the food you eat and avoiding shitty foods.

Shitty food isn't going to ruin your life. In fact, in the big picture, it's going to allow you to drop the fight, stop being so scared of food, and finally listen to what your body *really* wants. That may often be beautiful, artisanal farm-to-table hand-crafted food. And sometimes that will be delicious, neon, shitty food.

SHITTY DIET FOOD

Look, you are allowed to eat whatever you want, from now until the end of time. But I am going to make a case against diet food. Mostly: Why would you? We are trying to feed our bodies *actual* calories.

Diet foods reinforce the idea of eating "zero-calorie foods." They're created in a lab and marketed to people promising health and weight loss. And it's a scam. There's lots of evidence that when you eat fake sugar (aspartame, Splenda, etc.) your body still thinks it is getting sugar and releases insulin, but there is no sugar to use, so it's actually incredibly counterproductive. It can cause hypoglycemia, high stress hormones, etc.

I *could* argue that by definition, diet food has no place on The Fuck It Diet, because I am trying to lead you into a world where we embrace filling up with real calories instead of low-cal filler food. But this is still The Fuck It Diet, and if me telling you that you "shouldn't" drink diet soda is stressing you out, my god, just do whatever you want. However, I do urge you to again get honest with yourself about *why* you want to keep diet soda in your life. *Why* did you start drinking diet soda in the first place?

Some people claim they actually like the taste, and as an ex-diet soda drinker who used to drink maybe five diet sodas a day, I really question that. One of my students *swore* she really truly loved Fresca, only to realize that having it in the fridge was still a sign she was hoping to lose weight.

Guys . . . fake sugar tastes horrible. However, you're in charge. Do whatever the fuck you want.

PURITY DOESN'T EXIST

When I was fourteen, I was diagnosed with Polycystic Ovarian Syndrome (or "PCOS"), a hormonal and metabolic syndrome that's often associated with insulin resistance and weight gain. Basically, they don't know what causes it and don't have a cure, so they tell people to diet. My doctor told me to watch my carbs, exercise, and not gain weight. So I took my doctor's advice and I made a decision: *I can make this go away if I do this perfectly. I will lose weight, eat perfectly, and heal.*

I figured I needed to cleanse my body of carbs, cleanse my body of bad foods, and get rid of my obviously inherently unhealthy body fat.

And in case you're wondering how that worked, it was a total, decade-long disaster. It directly led to more bingeing and feeling *more* terrible about myself. I became obsessed with purity. And nothing was ever good enough—foods were never pure enough, and my body was never cleansed enough.

This is orthorexia: the obsession with food purity. It often happens alongside caloric restriction and weight obsession, but is its own miserable disorder. It is very common, and can easily pass as "healthy living." It's easy to trick *yourself* when you are orthorexic and convince yourself you just care about health. But obsession is never healthy.

Not sure if this describes you? Here's a quick barometer: if what you're doing is causing you stress, panic, or obsession, it's not worth it and it's not working. On The Fuck It Diet, eating shitty food is especially healing for people who have been obsessed with purifying their diet.

Allowing yourself to eat shitty food is like medicine for the mind. This is like cognitive behavioral therapy for orthorexics. You need to up your tolerance, and gain some neutrality around the kinds of food that used to scare you. Until you do that, your orthorexia will continue to have power over you from the shadows.

Being afraid of "shitty foods" (whatever *you* consider that to be) is just going to make your life more miserable than it needs to be. It's also impractical. You're going to be faced with shitty food out there in the big bad world, and unless you want to be filled with irrational fear at every turn, this is worth getting over. Purity doesn't actually exist, so thinking that way is misguided and based in control and fear.

If you have experienced a chronic health issue, I feel you. Some bodies *can* get overburdened by heavy metals and chemicals and whatever else, and different bodies have different weaknesses, but the purity model still doesn't work. It's just impossible. It's not how bodies or health or food should work. Instead, we have to reframe the way we look at health.

Back in the day, if someone put their impure corn chip in my pure guacamole, *they were ruining everything I had worked so hard to accomplish from my miserable, obsessive, pure, perfect eating. Because of all my toiling, my health was probably getting better very soon, but their impure corn chip was ruining everything.* That corn chip would lead to a breakdown and ensuing panic.

It was these kinds of all-or-nothing diets that also seemed to promise that once you fixed your body—through purity and perfect eating—you'd reach a point where you wouldn't be obsessed with food anymore. Purity would purify you. It would heal you. It would eliminate all your cravings, and any binges and appetites would vanish.

This does not work. This *should not* work. Because cravings and appetites are human, just like emotions are human, and in both cases, trying to suppress them will backfire hard-core, and make you more obsessed than ever.

YOU DON'T NEED TO GO ON A GREEN JUICE CLEANSE

My acupuncturist told me once: "Forget 'detox,' can you reframe it and look at health as 'supporting' your body? Or liver? Or hormones?"

We need to switch over to a different way of looking at health: *supporting and nourishing* the body, not purifying. You'll be way better off if you start supporting the body to work, circulate, and rest, so your natural detox system can do its thing.

Eat food that feels nourishing to you. Eat food that actually fills you up. Eat food that fulfills your cravings. Eat food that actually makes you feel good and satiated. Sometimes that'll be cookies, sure, but sometimes that'll be soup, sometimes that is pasta bolognese, or a spinach salad, and everything in between.

In a restrictive mind-set, the idea of "eliminating" certain foods or ingredients becomes disordered really quickly. The answer is to *add in* nourishment, not *take out* foods. Remember, your cravings are the perfect way to figure out what is going to nourish you today.

For instance, if you feel your digestion needs help, can you add in healing and probiotic foods instead of restricting? If you feel like you need or crave more greens, why not just add them to a meal instead of going on a juice cleanse?

The truth is, you can do whatever you want. You can eat however you want. You can add in or take out any foods you choose. You are ALWAYS in charge, day to day and meal to meal. If a certain food doesn't make you feel good, you have every right and every opportunity to try and avoid it or experiment with it. But I'm asking you to reframe what you've been taught about food and health. Start realizing that less is not always more. Sometimes less is actually . . . less.

You actually need lots of variety and calories to really support your health, your detoxing, your repairing, your hormonal health, your bones, your muscles, your brain, and your moving through this world. Start nourishing, and let it mean whatever it means to you.

SO MANY, MANY DIET RULES

Do you believe that doughnuts will go straight to your hips? And that eggs have too much cholesterol? Or that fruit has too much sugar? Or that you can't eat three hours before you go to bed? Or that gluten is *killing everyone*?

You need to take stock of what all those diets and diet books have done to you. Those rules are the reason you are now crazy with food. Those subconscious rules are still rattling around in the back of your brain. They are part of the reason you have binged. And they aren't allowed to be a part of your life anymore.

A client recently shared a connection she made between her

relationship to her food and her relationship to her life: "Thinking about food rules makes me think of 'rules' I've internalized from life . . . e.g., I've always been taught to finish what I started (clean my plate). I'm starting to think that developing a healthy mind-set around food will help develop a healthy mind-set around life. There is so much more overlap than I realized."

You will probably find that too. Once you start unraveling your food rules, you'll realize you also follow a million little rules in every area of your life. It's really helpful to become aware of what's in your subconscious and shine a light on it, so it can't rule you from the dark.

OLD DIET RULES

First make a list of all of the diets you have ever been on, no matter how short-lived.

Then I want you to list all of the different rules those diets implanted into your brain.

Then follow up with a list of "miscellaneous rules" that you soaked up from your aunts or the internet or yogurt commercials.

Then you can go through and write a counterargument for each one. Then you can burn it. Or you can be less dramatic and just write something at the end like, "Bye Felicia." And then burn it.

BUT WHAT ABOUT MY HEALTH?

You don't have to give up on health, but this book *is* asking you to redefine health, the way you seek it, and the role weight and food play in your life. I am asking you to consider that you may have been focusing on the wrong things in your quest to achieve health. I'm asking you to consider that what have been dubbed "weight-related diseases" may not actually be *caused* by weight but are more likely *stress*-related diseases. And dieting happens to be one of the most active ways we keep our body in a stressed state.

Having a dysfunctional relationship with food *is* unhealthy. Dieting and restriction and yo-yoing are not good for you long-term. So, on the most basic level, The Fuck It Diet is meant to improve not only your physical health but your overall mental, emotional, and spiritual health.

Health is so complicated and has so many nuanced contributing factors, but living your life in a suppressed metabolic state only hurts your health and your ability to connect to your body and true appetite. Lower weight does not automatically equal better health, and vice versa. Changing this association is a game changer. Remember that people at higher weights can be perfectly healthy, and that extra fat can even be *protective* when recovering from surgery. In fact, "overweight" BMI patients are less likely to die after heart operations and tend to live longer.[47] *Health* is listening to your body. And our *whole point* on The Fuck It Diet is to get you normal with food so you can eventually, *easily*, listen to what you want and need.

But listening to your body is next to impossible when your metabolism is impaired and you're still afraid of most foods. How can you truly listen and hear your body when your body is just screaming indiscriminately for more calories? How can you follow what it wants when you're judging 95 percent of the foods out there, and are more afraid of gaining weight than anything else? You just can't. Your eating will never be truly intuitive that way. So if what you want is health and to actually give your body what it needs, that lies on the *other* side of this whole journey.

Go on The Fuck It Diet *first*, and little by little your body will speak up more and more, and you'll be in a way better place to actually listen and seek whatever health you desire without all the stress, confusion, and obsession with food and skinniness.

Celeste told me, "In the beginning I only craved heavy foods, but one day I suddenly craved the cold crispness of a salad. I was so surprised by that. I have *never* craved salad before. In fact, I was *so sick of* salads I could barely choke them down. Now I find my cravings are very vast and varied, and I just sort of eat when I'm hungry. It's great!"

Beyond that, many people find improved health after taking away all eating rules. Carrie reached out to me to tell me about her health change after going on The Fuck It Diet. "I have been battling high cholesterol for months and was having to change meds and my diet constantly," she wrote. "Since starting TFID, I have been eating whatever I want regardless of its 'supposed' impact on my levels and today . . . they are the best they have ever been! I'm thrilled and I'm feeling amazing!!!!!!!!!" (Yes, she really added all of those exclamation points.) Of course, I am not saying that will be every person's experience, but sometimes the body just needs to be trusted.

Nobody knows the healthiest diet. I have had doctors and nutritionists give me completely conflicting diets to follow. One told me to eat paleo—high fat, lots of meat, low carb, no grain, no fruit, and absolutely no carbs *with* protein for digestion. One told me to eat lots of whole grains and lean proteins, stressing the importance of always eating carbs *with* protein for blood sugar. One told me to focus on fruits and vegetables and eat vegan. One told me to eat for my blood type and gluten-free.

Guys. It's absurd. If you tally up all the *bad* foods from those doctor-prescribed diets, *every food is bad.* And the reverse is also true. You will only be led into madness trying to reconcile all of the rules and advice out there.

If you are convinced that a certain diet is unequivocally the right one, there is a scientist who is just as fully convinced that the opposite is true. I've heard people claim that the vegan diet cured their hormonal imbalances, and people who have claimed that the vegan diet *caused* their hormonal imbalances. There is no ultimate diet truth, because *so* much more goes into our health—and the way we digest and process foods—than just the foods themselves. And to assume that there is one perfect diet for humans who have inhabited unbelievably varied terrain all over the world . . . *pshhh.* Let ultimate food truths go. Consider the alternative—that you need different things on different days, and during different phases of your life.

Health is far more complicated and holistic than a simple math equation. We have been led to believe that our weight and eating are the biggest factors determining our health, but it just isn't true. Melissa Fabello states, "Western Medicine hyper-medicalizes health—which seems sensible at first. But only because we've been socialized to believe that our bodies should operate like machinery and that with a little fine-tuning from doctors, we can live long and healthy lives. But no. Our health isn't only determined by what's going on in our physical bodies . . . so we need to think more broadly about it. Not because medicine isn't legitimate—but because it's limited."[48]

I also want to point out something that may seem a little controversial, but . . . you actually don't have to devote your life to total and complete health and healing. And I say that as someone who *has* devoted my life to healing. My body struggled—genetically, environmentally, emotionally, physically. And for so many years I dieted to try to make things better, and dieting made things worse. I blamed all of my health struggles on myself, but it really wasn't my fault, and when I ended up realizing that, it was so freeing. The amount of pressure and blame I put on myself for *not* being healthy enough was extra-debilitating. Not everyone will relate to struggling with chronic health issues, but if you do . . . I want to let you off the hook, too. You've tried. You care. But some things are really hard to figure out. Some things are not fully in our control. And some things pizza actually *can* heal. Life is very mysterious.

You don't *have* to figure it all out. And if mental health and quality of life are the prices to be paid for physical health, is it really health at all? Will it last?

For me, The Fuck It Diet was always just the next logical

step in my own journey toward well-rounded health. I had been so militant about my diet for so long that when I finally learned about what restriction does to your stress hormones and metabolism and mental health, plus the healing effects of demonized carbs and sugar, I suddenly understood how years of dieting had damaged my health instead of improving it.

At this point, I truly believe that stopping restriction is better for your mental health, physical health, metabolism, nutrient absorption, and inflammation levels. And not only that, but the body will speak up if you take time to feed it, inhabit it, and listen to it.

Know that if you feel too petrified to take a step forward, you can always find a nutritionist or dietician who is well versed in Health at Every Size and recovery from obsessive dieting or eating disorders. And be aware that many nutritionists have their own food issues and weight biases, so be sure to seek out a nutritionist who will help you, and not muddy this process even more. Your own cravings are king.

"BUT I'M A BINGER!?"

Sooooo many people think that their bingeing makes them the exception. They think, *Okay . . . I understand why people need to focus on eating if they're someone who has been restricting, but I am an OVER eater, not an under eater. I am mostly bingeing.*

But here's the truth: *most* people who diet are also bingers. I was a binger, and 95 percent of my readers and students over the years have been bingers. This book was actually written *for bingers*, not for people with anorexia. I like to think of eating disorders as a spectrum. Anorexia and bulimia are at the far end of the spectrum, and easy, neutral eating (our goal) is at the other end. The Fuck It Diet is meant to help the "disordered eaters" who are somewhere in the middle. Disordered eaters are the lifelong yo-yo dieters and bingers, all the way up to the casual dieter. That's not to say that this book can't also help people with anorexia and bulimia, but people recovering from those illnesses often need professional support. Again, if you are suffering any form of extreme restriction or self-harm, please, please, please seek help. You can find resources by visiting nationaleatingdisorders.org or by reaching out to their helpline at (800) 931-2237.

Here is the big question: Do people who are disordered eaters have eating disorders? Honestly, it just depends on how you define it, but if you ask my opinion, I would say a soft *yes*, because they are somewhere on the eating disorder *spectrum*.

But for people who identify as bingers, this is the whole problem: bingers think that the cure for bingeing is more control, without understanding how restriction is affecting them. Most bingers think they have binge eating disorder, and that bingeing is happening in a vacuum, and that just "dieting better" will somehow cure it—*but that is restriction*. For anyone worried that you have binge eating disorder, you probably don't. Stand-alone binge eating disorder is extremely rare, and usually the result of congenital disorders where you don't feel satiation, ever.[49] If you have ever restricted or dieted, by definition *you do not have BED*. Your bingeing is almost certainly a *reactive* disorder,[50] in response to past dieting or restriction. A body that binges after you put it on a diet is a body that wants to survive—even if the diet only lasts a half of a day.

My student Kim told me, "I used to think that I had this subconscious part of me that would always binge given the chance, that I needed to be on *constant alert* and *constantly vigilant*. I also believed that eating modern foods with addictive additives in them was also part of my bingeing problem. I mean, *for years* I was *pseudo* intuitive eating, pretending to listen to my hunger but always trying to curb my 'overeating.' After going through TFID, and allowing all foods, and stopping my *fear* of overeating . . . it isn't even close to a problem anymore. The whole problem was self-perpetuating. Even foods with additives in them have no power over me . . . it just doesn't matter. I'm amazed."

Bingeing is not survival mode gone wrong—it is actually survival mode gone *right*. But as long as we fight it, we are going to *feel* like something is terribly wrong. So The Fuck It Diet *definitely* applies to you if you are a binger. The body is in a reactive state that only eating and feeding yourself can cure.

WHAT TO *ACTUALLY* DO DURING A BINGE

If you are worried about emotional eating, you're not alone. Most people are convinced that they are emotional eaters, *or* that they have binge eating disorder. And they also often think that emotional eating and binge eating disorder are the same thing. We are still going to be talking way more about emotional eating *very soon*, but it is not the same as a binge. A binge is our manic, panicked reaction to restriction. It's the *I-could-devour-the-whole-world* kind of eating, and is the result of dieting and famine mode. But that doesn't help you if the binge is already happening. So what do you actually *do* if you are in the middle of a binge? What do you do during that horrible feeling of losing control, and stuffing your face, and spinning into that manic, rabid, compulsive food frenzy?

The answer is to reframe the whole thing.

First, stop resisting it. When you resist it, you are perpetuating the cycle and staying in the panic. *Oh NO I am BINGEING!? I am not supposed to be BINGEING!!??*

You are making it so much harder to step out of the overwhelmed, manic, compulsive cycle. Once you want the binge to stop, you are adding *more* pressure and mental restriction—which makes the binge *worse. I am doing something wrong. This isn't good!*

I know this seems like a silly mind game, but it's actually the paradox of stopping the resistance. Once you decide that what is happening isn't supposed to be happening, you'll freak out. As soon as you freak out, you're screwed. You're labeling and demonizing your natural response to restriction, instead of a more innocuous response: *Ah, look at this, it looks like I might be having an awesome midnight feast tonight. I wonder where I'm still restricting.*

You can even rename it. Call it whatever you want: A feast. Famine fixing. Eating yourself to the other side. It doesn't matter; all that matters is that you decide that what's happening is fine and even *helpful*. It's important that you remember that wanting to eat a lot is totally normal, totally important, and totally allowed. Then sit the fuck down, slow down, and enjoy it.

The point of slowing down isn't necessarily so you eat *less*, but instead so you surrender and actually allow yourself to eat. I am not going to tell you to chew your food slowly and take a sip of water in between each bite and to try and fill up on celery or whatever you used to do when you were bingeing. Because that is absurd, and arguably still a kind of restriction. Fuck that.

I'm telling you to sit down, and slow down, and enjoy it, to *prove* to yourself that you are allowed to eat. Always. You are allowed to eat *just as much* as you would be eating if it were still considered a binge—you're just going to do it slowly to remember that you are allowed to be doing this.

Seriously. The goal here is NOT to stop eating, but instead, to eat, because eating is the only way you will ever get to a place where you *can* eat and not feel out of control. When you allow yourself to sit there, truly eating, not rushing to be finished, not stuffing your face because "tomorrow you're not allowed to eat this way," and not thinking *Why am I still DOING this?!?!* . . . I promise you, it will be a game changer. In this brave new world, your hunger is your friend, so fucking eat.

DOESN'T THE FUCK IT DIET CAUSE OBESITY???

If we truly want to address the "obesity epidemic," respect is the way to do it.

—MICHELLE ALLISON, THE FAT NUTRITIONIST

People afraid of gaining weight *always* ask, "Well, if eating is the answer, then how *do* people become three hundred, four hundred plus pounds in the first place? Isn't obesity caused by a version of The Fuck It Diet?"

No. Weight set ranges are complicated and complex and are the result of many factors. In many cases a person's weight range is genetic, sometimes inherited from ancestors surviving famines. It can also be the symptom of hormone, thyroid, or more systemic imbalances, which most often are environmental and genetic, and which food restriction and shame will never heal. Rising weight set points are also arguably caused by *under*nourishment and suppressed metabolism, which is perpetuated by dieting and restriction, and the yo-yo that ensues.

Most importantly, weight set ranges are more powerful than your own attempts at willpower. So fighting your weight set range is only going to cause more misery and hunger. It's important to remember that dieting is the surest way to mess up your metabolism and continue to raise your weight set range.[51] That's the big tragedy here: the very thing we are told to do to try to "save ourselves" is just making everything spin more and more out of control.

Also, the words *obesity* and *overweight* are also part of the shame problem. Trying to scare and shame people into losing weight isn't just ineffective for weight loss—it's damaging to people's relationship with food, their own bodies and feelings of worth, *and their health.* Weight stigmatism and discrimination doesn't put a fire under people to become healthier, thinner version of themselves; instead it leads to weight cycling, eating disorders, emotional pain, and lots and lots of stress, which breaks down health over time.[52]

Our ultimate goal is to fix our dysfunctional, addictive relationship with food. I know it's against everything you've been told up until this point, but the mainstream way of dieting doesn't work for *any of the things we want it to.* It doesn't work with our eating, and it doesn't work with our weight. So consider the radical idea that our bodies are actually trustworthy when it comes to appetite and weight regulation. We actually just have to get out of the way.

"BUT REALLY, WHAT IF I HAVE HEALTH PROBLEMS?"

We already know that dieting is not the answer for weight, long-term. Despite what we have been told, and despite what seems like common knowledge and common sense, dieting actually often does the opposite of what it tells us it will do. Are you willing to open yourself up to the possibility that dieting is equally fruitless for other health issues?

Are you really sure that your particular health problem requires a specific diet? I ask people to get very honest with themselves. Unless you have celiac disease, diabetes, a serious peanut allergy, or something of a similar extremity, a self-imposed diet (and in my case, even a doctor-suggested diet) may be completely unnecessary, unhelpful, and make it even harder to actually eat in a way that is healthy and intuitive *for you.*

If dairy upsets your stomach, I am not saying that you should ignore that and force yourself to gorge on milk. Nobody should try to sustain themselves on food that makes them feel horrible. But I *am* saying that the manic, perfectionist, obsessive fear of certain foods is keeping us in a vicious cycle and ends up giving that food way too much power, and potentially giving you stress-related health problems at the same time. Instead, if dairy was allowed and neutral, you would do a way better job of listening to what your body wanted and needed to feel good. (*And* you'd avoid the possibility of digestion issues just from the fear and stress itself.)

My student Molly is a plus-size yoga instructor who'd been wanting to get pregnant, but was having a lot of trouble, for a long time. She was told by her doctors that in order to get pregnant, she needed to avoid carbs and lose weight. She is fat and has PCOS, so her hormonal issues were blamed on her weight and her "imperfect" eating. So she would follow their advice, diet, and lose some weight (miserably and obsessively, of course). She was eating *exactly* what her doctors told her to eat, and she *still* wasn't getting pregnant. So of course she blamed it all on herself, and assumed she just wasn't doing it well enough—and that's why she hadn't lost "enough" weight.

Eventually, she went on The Fuck It Diet, and she also started supporting her insulin response with certain supplements, recommended by her nutritionist. We started working together, and she learned to eat what she wanted—and stopped trying to lose weight. And guess who got pregnant and now has a healthy one-year-old baby *without* ever sticking to the doctor-prescribed low-carb diet, or losing weight?

"Learning to love and trust my body has been a process. . . but when I look at my sweet baby girl, I'm reminded that my worth truly doesn't come from my appearance or my weight or the perfect 'clean' foods. I am happy to be able to teach her the truth. I hope that if someone ever pressures her to go on a diet, she will say with confidence, 'FUCK THAT!'"

Of course, everyone's experience will be different, but these stories of healing are really common. For instance, Bridget used to think she couldn't eat dairy, only to realize that she *made* herself intolerant of it when she weaned herself off it. Now she eats dairy every day again and feels great.

And Meredith was told by her doctor to eat a low FODMAP diet to manage her irritable bowel syndrome. FODMAP is an acronym for different molecules found in many, many foods that can be poorly absorbed by some people. So, for nearly two years she ate a very limited diet. She limited wheat, sugar, certain fruits and veggies, dairy, and a lot more. "Now I realize I can eat pretty much anything I want, and that in reality, it was just my digestive system healing from an eating disorder and a very nasty GI bug, combined with lots of stress."

Kara just reached out to tell me, "I struggled with 'gut issues' for a long time. I tried elimination diets and was on about four different medications to control my symptoms. But now I'm in remission! Know how I did it? EATING FOOD. More food, enough food, every day. Amazing, huh?"

Food intolerances *do* exist, of course, and you should never force yourself to eat foods if they cause pain or discomfort. That's the beauty of The Fuck It Diet. You don't have to eat anything that makes you feel bad—either because you just don't want it, or because it doesn't make you feel good, like giving you an upset stomach or a headache. I am, however, giving you the *permission* to take away the guilt and fear. You don't need to eat perfectly. And in time, you may just find you can handle more than you thought.

For those of you with serious food allergies or diabetes, you can still learn to listen to your cravings and your hunger *within* your necessary food limitations. And you can and should always consult a health-care provider you trust. I have a resource list of Health at Every Size practitioners at thefuckitdiet.com/resources.

One of my clients is married to a surgeon who often tells patients to go lose weight before he can treat them. Recently she asked him how *often* people actually lost the weight and came back to be treated, and his answer was, "Well . . . never." Never. That means that they likely *tried* to lose weight, failed, and so they never returned, but probably kept trying and began a weight yo-yo that, as we know now, is terrible for your health—*or* they actually found a doctor who would treat them as they are.

Unfortunately, it's incredibly common for doctors to be stuck in deep weight bias. Instead of helping fat patients the way they would help a thinner patient, they insist that they can't do anything until the patient loses weight. Fat patients are often told their health issues are their fault, and that their weight is the root cause of every complaint, even if their weight has nothing to do with, or is just a side effect of, their genetic joint problems, hormonal imbalances, or chronic illness.[53]

If you find yourself stuck with a doctor who is too focused on weight, it can be helpful to ask how they would treat a thinner person with the same health issue, and insist on being treated that way. Or find another doctor. It's in your best interest in the short and long term to find a doctor who will support your health without blaming everything on your weight.

If you are nervous about your health, just remember, you aren't going totally rogue. You're actually just learning to eat. You're learning that you don't need to avoid a food that a magazine told you was bad for you, and challenging the assumption that weight is the *cause* of all of your health problems.

FIND A WEIGHT-NEUTRAL DOCTOR

If you are afraid to go to the doctor or feel like your doctor isn't aligned with your new way of relating to your body, you deserve a doctor who will work for and with you, whether you're fat or thin. Try to find a doctor who will give you the same options no matter what your weight is. You are allowed to seek health without being shamed for your weight or eating.

» TOOL #2: LIE DOWN

I don't want to "lean in"—I want to *lie down.*

—ALI WONG

Yep. Really. Lie down.

Every day, take at least ten minutes and lie down. After work. During lunch. While your kids are napping, or at school, or at soccer practice. Lie down. Lie down in your bed. Lie down on the couch. Lie down on the floor. Lie down on a yoga mat. I don't care where, just lie down, close your eyes, and do nothing for ten minutes.

You can get nice and comfortable, you can use pillows and blankets and eye pillows if you want to. But do nothing. Just let yourself be, and do nothing, for ten minutes.

You'll be tempted to bring your phone into bed and browse. Don't. Not looking at your phone is the one part of these ten minutes that is different from everything else you do all day long: I need you to give yourself a moment to *stop* doing. Stop having an agenda. Stop taking in info. Stop trying to figure things out. Stop scrolling.

Your brain won't stop going—it never ever will. That's okay, that's just what brains do. Just give it permission to stop if it wants to. Give it a *chance* to rest, even if it doesn't take you up on the opportunity.

We never ever let ourselves rest. We don't think we deserve it. We don't think we are allowed, physically or mentally or emotionally, to take the time for ourselves. We don't think it's worth it or helpful. But because we never slow down, our stress hormones are always pumping at high gear. Physically, these stress hormones actually increase our risk of disease.[54] And mentally, this puts us in chronic fight-or-flight mode, constantly on alert. We think that pushing a little harder is going to be the answer. It's not. Lie down.

The people who do not believe they deserve to lie down and do nothing for ten minutes really, really, need to lie down and do nothing for ten minutes.

The people who do not believe they have the *time* to lie down and do nothing for ten minutes *really, really, need to lie down and do nothing for ten minutes.*

My student Chiara told me, "I was resistant and skeptical, but I cannot believe how much doing the lie-down has changed my life. It is so simple, but so restorative. It definitely makes me calmer, but I think I'm now a better, kinder person too."

This is *physical* self-care, giving your body a genuine, adult time-out. Prove to yourself that you deserve ten minutes to lie horizontal.

And, hey, if you want to start doing daily two-hour lie-downs, I would never, ever dream of stopping you.

THE NOBLE ART OF REST

Slowly but surely, as I have taken myself through my own Fuck It Diet, I have realized how essential *rest* is. Not only for the physical part of this process but also for the mental, emotional, and more existential/symbolic parts of this journey. If you resist rest, as many are tempted to do, just know that rest must have its way with you, whether that is months or even years down the line. This whole thing can't fully work without it. The ten-minute lie-down is *just the tip of the iceberg* on our rest journey.

This is good news, **because rest is awesome**. But it is also terrifying to so many of us who subconsciously only feel worthy when we are constantly working or being productive. Rest is the cure for the workaholic. It's the cure for the nonstop, constant productivity that often accompanies eating and body-image issues.

On a very physical level, dieting puts you into a state of constant adrenaline and cortisol. Your survival mode is *on*, and this accounts for some of the euphoria the body can feel during caloric restriction. This also accounts for the body's ability to override not eating enough calories, not listening to its cries for rest, over-exercising, overworking, overworrying, and not being comfortable sitting still or relaxing.

This survival state can be very helpful in *real* times of struggle (war, famine, being hunted by a lion . . . i.e., major crises). It will save lives in dire times, but it is not sustainable long-term. And pretty plainly . . . it is no way to live. Our subconscious fears of never doing enough and never being thin or beautiful enough are not allowed to keep running the show.

Over time this stress state will break your body down. It will deplete your adrenals and wreak havoc on hormones, and to come out of this state will require major . . . ***rest***. And food. And rest. As well as a major overhaul in how you look at life.

Besides all of these physical ramifications, we also rarely let ourselves mentally or emotionally rest, which, in turn, affects our *physical* bodies. Even a low-grade feeling that you're never accomplishing enough, or going fast enough, can keep your body in a state of depleting, low-grade anxiety—with high stress hormones. Your mind affects your body, your body affects your mind, and round and round you go in your hamster wheel of stress, never quite sure where it is all stemming from.

Even people who would never consider themselves type A or workaholic would be surprised how many *stressful* beliefs they have about rest, relaxation, and taking *real* downtime to just be with themselves and get nothing done. We just don't think we deserve it. And we *definitely* don't think we deserve it in an imperfect body. But that, my friend, will keep you in a world of exhaustion and misery for the *rest* of your life. No pun intended. I never intend puns.

There are two sides to our autonomic nervous system—sympathetic and parasympathetic—and they balance each other out. The sympathetic nervous system is concerned with short-term survival: it keeps your basic organs running, keeps you breathing, and runs your fight-or-flight mode. The parasympathetic runs the "rest-and-digest" or "feed-and-breed" mode, the calmer side, and this side is all connected by one big nerve.

Fight-or-flight is a high-alert state run by stress hormones and is activated from crisis—think famine, danger, trauma, and fear. Being stuck in crisis mode long-term causes inflammation and low metabolism, and just general depletion.

We want to turn off chronic fight-or-flight mode and start living in the other mode: the rest-and-digest and feed-and-breed mode. And the way to do that is to rest. To take time-outs. To eat. And to *breathe*. Breathing actually *physically* helps activate the parasympathetic feed/breed nerve that connects your heart, your lungs, and your digestion. Activating this nerve will calm you, lower your stress hormones, help you digest, and get you out of high-alert mode, which is basically the all-around underlying *physical* goal on The Fuck It Diet: to chill out.

Rest is a huge theme throughout this entire book. We will talk about this even more when we get to the emotional part and can focus on breathing and feeling. But getting out of stress and crisis mode isn't just a psychological or emotional challenge—there is a very important physical component. You're now going to be giving yourself a small chunk of rest in the middle of every day with the lie-down. But let this be the bare minimum of rest you give yourself.

You must start carving out very real chunks of time to do nothing. Your body will most likely get more tired before it starts to come back to a more repaired state, especially if right now you are still in a wired and adrenaline/cortisol "go go go" state. This rejuvenation may take months. And it may take even longer than that. That is okay.

I know life is exhausting. I know that having nonstop jobs, and kids, and never-ending family obligations, make it seem impossible to rest. But at the same time, you need rest more than anyone else. Can you let yourself off the hook? Can you take a mental health day? Can you turn down an invitation you don't have the energy for? Can you ignore a chore that doesn't need to happen today, and take on this new, deeply healing chore? ***Resting* is your new chore.**

My student Meredith shared her journey of adding in more rest with me recently. "In the beginning my inner asshole would call me things like 'lazy' and 'unproductive.' But I have been allowing more and more rest. Just realizing how much rest your body needs to heal, especially when you're dealing with emotional healing, was life-changing. Now I lean into the rest real hard. I let myself sleep in later every chance I can, I just ordered a weighted blanket to make my resting even better, and I even go to yoga classes where you sit in restorative poses for two hours while someone gives you a Thai massage. This rest thing is incredible! I'm a convert!"

Mark shared his experience of taking off the pressure to be productive. "I went through a few months of feeling unmotivated, and I'm glad I chose rest more times than not. Eventually, with enough rest, I started to want to do other things again. It happened on its own, once I gave in to rest."

We have beliefs about rest, either that we don't have time, or that rest isn't very helpful or important—but those are wrong. Tim Ferris interviewed successful people on their most important practices for his book *Tools of the Titans*, and one of the most consistent overlaps that all of these billionaires, icons, and world-class performers had in common . . . was prioritizing rest. Because rest breeds *sustainable* productivity and creativity. We need those times to lie fallow. We cannot live in a perpetual harvest!

My friend Emma said to me, "I find when I prioritize rest—

and fit in work *in between* resting—I am paradoxically way smarter and more productive."

Just like all good things, you will probably find a way to resist the rest piece of this journey, and I will be here to remind you how important it is.

REST

In addition to your ten-minute rest sessions, I want you to schedule in some couple-hour chunks to "do nothing" this week. And when I say "do nothing," I mean allow yourself solid chunks of time where you don't need to do anything productive. Schedule yourself frivolous downtime. A nap. A chunk of time to catch up on TV. Window-shopping. This can be your personal brand of frivolous downtime. In sweats or out in your Sunday best. It doesn't matter what you do, but can you try to experience frivolous downtime and not let yourself feel guilty about it at all? Could you even begin to see it as healing?

(We will be exploring all the mental hooks that our society has in our subconscious about rest in later phases, but for now, start with the simple physical act of rest.)

WHAT ABOUT EXERCISE?

Exercise is really good for you *if you are fed and rested.* Exercise strengthens you, circulates your blood and oxygen and lymph system, and is incredibly life-affirming. But exercise is *not* healthy when you are in a famine state, or when you are depleted or exhausted. Too much movement and exercise is *just* as damaging to the metabolism as dieting and restricting. It is the *other* side of restricting.

Why is excess exercise and cardio unhealthy? Think about it this way: if you were, say . . . being chased by a lion for 5, 10, or 26.2 miles, the body would worry for your life, put you into an adrenaline state, and hold on to weight. *Will there be more lions chasing me? Will I not be able to feed myself at the rate I am expending energy?* Cue conservation mode. And also, cue adrenaline survival mode. If you overexercise, *especially* cardio, you are hemorrhaging energy. You need a shit ton of food if you are doing a shit ton of exercise to avoid these survival modes.

We are designed to eat, move enough to do whatever our work is, and rest. *Rest* is a big part of how we are biologically intended to spend our days.[55] That's *thrive* mode. The idea of needing to constantly be moving is completely insane, a sign of crisis, and can mimic a life-threatening situation. There is simply no reason to run at full throttle for that long, and our bodies respond accordingly.

What that means is, if you are tired, you need to rest. And at the beginning of The Fuck It Diet, you may be tired for a good few months or longer. Remember how normal that would be, if you were recovering from a famine?

I cannot predict your own personal rest needs, but as long as you are tired and feel like lounging . . . lounge. Light walking and yoga? Only if you feel like it. Let yourself rest as long as your body needs it and let yourself move only in ways that feel good. You do not need to ever do another kind of exercise that you don't like. And you do not *ever* need to exercise when you are tired. So no more sneaking in a run right before bed when you're already tired. No more dragging yourself to the gym at 4:30 a.m. when you're so tired you might fall off the treadmill. This is the beginning of a lifelong relationship with exercise that *enhances* your life instead of punishing you or keeping you running away from your exhaustion or your feelings. This is the beginning of feeling out, day to day, week to week, *Am I rested and fed enough to exercise?*

I just want you to understand how damaging it is to force exercise on an impaired metabolism. We automatically conserve energy when we are expending too much (exhaustion, lower metabolism), so the idea of calories in versus calories out simply doesn't work the way it would if we were machines. Just like restricting and dieting, overexercising is actually doing the *opposite* of what you hoped it would. It is going to keep you in a vicious cycle, stuck at the beginning of The Fuck It Diet, and barely experiencing any of the benefits.

For some people, being told to exercise less and rest more is the best thing they've ever heard, and for some it's the worst. So especially for all you exercise "junkies" who panic at the idea of resting—start with the lie-down and other lie-down activities, add in rest where you can, and reframe exercise as something that is rest-dependent.

My student Maura used to be a workaholic and overexerciser, as well as a dieter. She told me, "Allowing myself to rest, and not forcing myself to move at all, totally helped heal my relationship with movement. My body eventually began to crave very specific types of movement, and I actually enjoy it now. But first I just had to rest and listen, and eventually my body sorted itself out."

Harriet said, "I had such a negative relationship to exercise because I forced myself into it too many times during my disordered stage. I used it entirely as a form of punishment and control. After letting myself off the hook, and resting for a while, I do often feel the need to move and have been working to honor that more. But sometimes I just allow that to be stretching. I never force it, and I consider the fact that I can even hear and feel what my body craves to be a great healing step in the right direction."

In the beginning of The Fuck It Diet, assume that you need to rest. And when you finally crave movement again, *reframe* the way you look at it, and look for strengthening and stretching and circulating over "burning calories."

I know it may still feel terrifying and counterintuitive, but stopping your intense exercise for a while is the only way to really let your body and metabolism heal.

PERMISSION TO NOT EXERCISE

Take this entire week and give yourself full permission not to exercise. Allow yourself to lounge. If you feel the genuine urge to go for a walk, honor that. Otherwise, VEG OUT.

This exercise in nonexercise should continue for as many weeks as you want. The goal is to follow genuine impulse and desire, not fear and compulsion.

HOW WILL YOU KNOW THIS IS WORKING?

There are a few big things that will happen that will be clues that everything is going in the right direction:

- » You'll start forgetting about treats you bought that you would have previously inhaled or obsessed over.
- » You'll start actually noticing whether you like the taste of certain foods or not.
- » You'll notice that sometimes you're in the mood for certain foods and sometimes you're not.
- » You'll feel comfortable stopping in the middle of a certain food if you're full or if it isn't the food you really want.

- You'll stop worrying that eating certain foods will directly affect your weight.
- You may even get strangely picky, and not be in the mood for foods that you used to love. You may be bored by lots of food. This is normal! And this is also a sign that things are shifting.

Here are what some of my readers and students have said when I asked when they knew The Fuck It Diet was working for them:

I knew this was working when I put down a partially eaten cookie because it didn't really taste that great. That is something I previously wouldn't notice until I'd eaten the whole thing.

I knew this was working when I was surrounded by sweets and none of it looked good. My body wanted tangerines.

I knew this was working when a pint of Ben & Jerry's started taking me a couple of weeks to eat instead of eating the whole thing in one sitting.

I started eating only 2–4 cookies in a sitting instead of half the box. And these things happened without any effort or control. It wasn't me forcing smaller portions, it just happened by allowing all food.

I still crave sweet things, but I have no desire to eat ALL of them, just the high-quality desserts.

I have no idea how many calories I eat each day and I'm so happy for it! That is pretty significant for me because I used to audit calories like I was balancing a bank register every day.

My colleague brought cake with her and my body didn't even want a slice of it and I didn't waste a thought on it. The cake was next to my desk at work all day and I was perfectly fine. In the past I would have thought about it all day and eaten it all in the end.

I used to feel this magnetic pull on my mind AND my body around food. And now I'm like "Eh, I don't feel like anything sweet right now." Until of course, I do, at another time of day.

I keep forgetting that I have chocolate or cookies in my pantry. That would never have happened before The Fuck It Diet.

I stopped eating chocolates while there were still some left in the box!

I'm just less excited about eating. I still love to eat but it is less obsessive now.

I'm able to have old trigger foods in the house without thinking about it nonstop before eating until it made me sick. I still crave certain foods, but don't feel as out of control anymore.

And this one comment summed up what happens on The Fuck It Diet:

> *Basically, being able to eat as much as I wanted of everything made me not want everything.*

It's really, really important to note that none of these people experienced any relief from their food fixation by "trying to stop fixating on food." They'd all tried that before.

Willpower was NOT the answer to any of their issues with food. Willpower was the thing that kept failing over and over and over. What finally worked was allowing and *actually eating* whatever they wanted. That is the paradox of The Fuck It Diet. And as much as I *love* the mystical, this is not mystical at all, it's biology.

It's also important to note that there is *no set time frame* for any part of The Fuck It Diet. There is a *reason* why this isn't broken down into weeks. There is a reason why I didn't promise a ninety-day fix. Because this journey will take different amounts of time for different people. And putting pressure on yourself to heal in a certain amount of time is just going to put you into a sort of antidiet pressure cooker that antidiets don't belong in. So *please* remind yourself that you have all of the time in the world.

You don't need to get anywhere fast. All you need to do is what you can do, and trust that it'll unfold in the right timing. Quick fixes don't tend to stick anyway.

JUST ONE CHANGE

In the thick of it, it can be really hard to tell if anything is improving, but it can be helpful to deliberately take stock of anything that has changed or shifted. Anything at all. Has any food lost its power over you? Has just the prospect of not having to diet or lose weight made you feel lighter? Has the concept that you can accept yourself and be healthy at any weight been liberating? Are you able to enjoy any foods that you couldn't before? Anything.

Write what has changed or shifted since starting The Fuck It Diet. No matter how small, that change means it's working. Slowly, but surely, it is working.

YOUR BODY IS FREAKING *SMART*

Look at it this way: During all those years of bingeing and feeling like your body was betraying you, your body actually knew *exactly* what it was doing. It was trying to re-feed and heal your metabolism so you could go about your life. Yet we assumed we were smarter than our bodies.

All those years of mistrusting our bodies and trusting health gurus, and our bodies just wanted to eat. They wanted to eat exactly the thing that would help them find balance.

So what I want you to do is allow yourself to trust that your body is right. Trust that your body has your best interests at heart. Meaning, if you are tired, you need to rest. Not push through. If you are hungry, you need to eat, not push through. If you are sad, you need to cry or take time for yourself. If you want potato chips, there is probably a pretty good reason, and you should follow that.

Your body has always been smarter than you. Your body works on instinct and intuition, both of which have access to really profound information. Your body knows when you need to eat, when you need to sleep, *what* you need to eat, and even knows when you aren't on the right path. Your body is where the wisdom is. Trust it.

THE EMOTIONAL PART

Now that you're feeding yourself and resting, it's time to move on to feeling. This is the part that some people will think is pointless or unrelated. So let me explain up front why it's in here and what you will get by actually using the tools and exercises in this section devoted to emotions and feeling.

First, emotional eating will naturally become a nonissue once you start getting in the habit of feeling instead of avoiding your feelings. The physical part (eating more, paying attention to your hunger, and listening to your cravings) will help heal the yo-yo bingeing cycle, and *this* part, the emotional part, will naturally fix our tendency to numb or avoid our emotions with food. Pretty soon you'll be enjoying brownies, not because you're denied *or* stressed, but because you fucking want brownies, and then move on with your life.

Because of that, the other thing you'll start to notice as you work through the emotional part of this journey is that the panic and insecurity that accompany your relationship to your body and weight will have way less of a hold on you. It won't feel so overwhelming.

Also, being willing to feel, as opposed to suppressing your

emotions, will help your body to live more in a *thriving* and calm mode. Breathing and feeling directly help your body activate that feed-and-breed, rest-and-digest mode of the nervous system, which supports your physical health, sleep, and willingness and ability to face things head-on. *It's way better on this side.*

At this point, if you've been applying the lessons from the physical section, you are probably bumping up against a lot of old mental and emotional stuff that doesn't feel so great. We have a lot of emotions and a lot of beliefs about weight, food, and the way we look at ourselves. It can all be really loaded and overwhelming.

As you might imagine, the mental and emotional processes are very closely connected. Your thoughts and feelings are tangled up in each other and constantly affecting each other. It's jumbled. But the way I teach it to you can't be as jumbled, so remember I've separated this book into different parts. It'll be helpful to learn about feeling and thinking *separately*, before we put them together in practice.

The mental part will be all about what you *believe* and think, and this current emotional section is about the emotions you *feel*. Or maybe, more importantly, what you have been avoiding feeling.

EMOTIONAL EATING VERSUS BINGEING

Lots of us believe that we eat our feelings and *need to get some control goddammit*. But what is actually even more common, and way more detrimental, is using *dieting and control* to numb our feelings.

Dieting is one of the big ways we try to avoid feeling our bodies. It is a perfect storm of distraction, control, perfectionism, and the chemical high we get from adrenaline and other stress hormones when we restrict our food.[56] Dieting is a way to disconnect us from our bodies and stifle our life force. *Less* food and *higher* stress hormones help us temporarily live with less feeling, not to mention all of the concentration and focus it takes to override your appetite and famine response. Dieting is a distraction.

All of the following tools and exercises are devoted to help you begin feeling. You are going to be doing enough work on habitually feeling emotions, and when you do that, emotional eating will lose its power. When you are committed to feeling emotions, the coping mechanisms you've used naturally begin to take a healthier role in your life.

We become dysfunctional with food, and it's not usually because of emotional eating at all. It's because of restriction, guilt, and the biological survival cycle that it induces—and also because of our deep fear of weight and taking up space. As long as you are getting out of reactive famine mode, *and* learning to feel instead of numb, you don't need to worry about emotional eating.

Do not get stuck in the trap of worrying that you need to stop emotionally eating—that will become restriction. Just eat and feel. I also want to remind you that eating is never a problem in and of itself. Shift your attention instead to making sure you aren't trying to numb or distract yourself with control, perfectionism, and *dieting*.

It's also important to note that emotional eating and bingeing aren't the same thing. In all of my experience working with so many, manyyyy people who believed that emotional eating was their problem . . . once they stopped restricting, they were able to see that emotional eating was not actually their *core* problem at all. Emotional eating is not the reason most people are so dysfunctional with food. The problem is the biological famine response and the yo-yo of restriction, which leads to bingeing. That is the thing to fix if you want to become normal with food.

Jenny told me,

> I used to think that emotional eating was my issue, and now I'm seeing how much it was tied to restriction. Now that I've gotten out of the restriction yo-yo, I'm not even sure I'm noticing any emotional eating anymore—it's sort of mixed in with just feeding myself. If I'm having a particularly stressful day, I might feed myself something a little more comforting, but it doesn't feel out of control or "bingey." I still cannot believe that I have a pint of Häagen-Dazs in my freezer and haven't finished it yet. I did not understand how people did that—now I'm becoming one of those people . . . mind-boggling.

Even if you worry that you tend to eat to numb, the answer is still never to restrict, because that will just start up the yo-yo again. In the physical part of the process (and always) the answer is eating. The answer now is also feeling.

The truth is, eating emotionally is actually perfectly healthy and normal. All humans eat emotionally. Your cravings and your body's needs are directly affected by your mental state and your stress levels. You are *meant* to have the option to comfort yourself with food. You are not a robot eating battery pellets for energy.

Food is fuel and nutrition, but it is also allowed to be comforting and grounding. And eating food to comfort and fill you up or to connect with others is a *nonissue* when you're not in a reactive binge/repent cycle. When you are actually feeding yourself and trusting your body, *and* your body actually trusts that it is being fed, comfort eating can just be a part of a very normal relationship with food.

For example, eating birthday cake is an emotional reason to eat food: celebration. And eating a big bowl of macaroni and cheese when you are tired and sad is a legitimate way to comfort and feed yourself.

Emotional eating will happen. It is part of being human. It is part of being a normal eater. It is part of assessing what we need in any given moment and trying to comfort ourselves. That will always be okay. The more neutral food becomes, the more your body will naturally, without thinking about it, balance and account for times when we eat emotionally or "more than we need." That's what a healthy balanced body and appetite does. You don't need to do *anything* about it. Just eat. Just listen. Just trust that this eating thing *ain't that deep.*

Emotional eating *is not bingeing.* The problem only comes when you feel guilty about emotional eating. *That* triggers the guilt-and-repent cycle: Feeling guilty for eating. Deciding to restrict a little to make up for it. Then all hell breaks loose. Then you're back in the yo-yo. The bottom line is, don't feel guilty about eating, because that will only perpetuate your dysfunction with food.

THE EMOTIONS WE AVOID

We use all sorts of control issues, perfectionism, and workaholism to constantly distract ourselves from feeling. These are coping mechanisms to help us deal with a life that feels too out of control and too painful. And humans are masters at avoiding uncomfortable feelings.

But our habit of suppressing and avoiding emotions backfires big-time, just like dieting. Avoiding the natural process of feeling emotions and being in our bodies keeps us in a cycle where we'll seek out anything and everything to numb or distract us from what our feelings feel like.

In order to feel our emotions, we have to make the journey from the mind into the body. We spend so much time up in our minds, and rarely feel what it feels like in our bodies—but emotions do not happen up in our minds. Emotions are energy moving through our bodies, but they can be uncomfortable, so we do whatever we can to avoid feeling them. But when we don't feel our emotions, they end up having *physical manifestations* in our bodies: tense muscles, a tight upset stomach, and even back pain have been linked to unfelt emotions.[57] When we get used to avoiding our emotions, the emotions don't just go away, they manifest in our bodies, waiting to be felt and processed. But we still keep trying to avoid feeling them, so we often stop spending time in our bodies altogether.

Nothing good comes from avoiding being in our bodies. All healing of physical ailments, emotional turmoil, and trauma . . . it all has to happen in the body. The more we avoid our emotions, the more power those emotions have over us (sound familiar? Same thing as *food*, maybe?). *No wonder we're so miserable.*

But most of us would rather do anything than feel our emotions, so we develop different coping mechanisms to help us "get out of our bodies" so we don't have to feel. We do anything we can to avoid feeling the fear, the pain, the sadness, the anger, the jealousy, and sometimes even the happiness. We just don't want to feel it. It's too much.

Allowing yourself to start feeling what's in your body may not seem very fun, but it will turn out to be soooo worth it. Once feeling and dealing with emotions is second nature to you, being in your body becomes so much easier. When in doubt, *feel.*

WHAT HAVE I DONE TO DISTRACT MYSELF?

Write stream of consciousness for ten minutes about what you may be using to distract yourself from feeling. Just write whatever comes to mind. And remember, these activities aren't the problem. You don't need to try and stop everything that comes up—all you need to do is just start to have awareness.

WE HAVE SOOOOO MANY EMOTIONS ABOUT OUR WEIGHT

The Fuck It Diet is going to bring up emotions—new and old. It's going to bring up panic, insecurity, pain, and memories of old experiences that led you to start dieting in the first place. It will bring things up that you'll want to shut down and lock away so you never have to feel. You will be tempted to use any kind of coping mechanism to convince yourself you have some control over the situation, and so here is my opportunity to tell you that it won't help, and that we need a new way of dealing with our emotions.

Joy wrote to me,

> I had such an emotional reaction to the idea of gaining weight, I really halted this process for a long time. I didn't think I was ever going to be able to just trust and accept my body and what it needed to experience as part of healing. So, I would waffle (no pun intended), I'd panic, and try to briefly go on a diet again, thinking that losing some weight would make me feel calmer, saner, and happier. It didn't. It's only when I realized that I had to really surrender to this, and all of the emotions about weight (and everything else) that I had been so petrified to feel, that I started experiencing an amazing new freedom. It is conceptually scary. And in actuality, it is scary too. But it is worth it, and I am so glad I didn't give up.

Lots of people want to quit this *whole thing* when things get scary. And lots of people *do* quit. If you choose to quit, that is your prerogative. Run for the hills, back into the arms of the diets you wish loved you. Run back to the safety of trying to control your weight. I will understand if you do. But then if you want to come back and keep going, I promise it is worth it.

I would have times in the beginning when I'd freak out, especially times I needed to go to "an event" and see people I used to know. I would temporarily forget everything I knew and cared about, and panic that my shirt *looks horrible what was I thinking I am trapped in this body with these boobs that no shirt fits*. I was having a meltdown over letting go of the things that I used to *use* to try and feel safe and worthy. The answer, in those moments, was feeling. Actually dropping my awareness into my body, and feeling the discomfort and fear that was coursing through my body. We think that feeling will destroy us, but learning how to move through the emotions that come up by feeling them and honoring their existence leads to more peace. Emotions aren't a reason to quit, they are just a reason to feel.

Your new task is to let yourself feel and honor this panic and these emotions. You will have emotions about the prospect of gaining weight, you will have emotions about your body, and emotions about making this big change in the way you relate to yourself, and weight, and food, and *worth*. You will even have emotions about the prospect of having emotions. And even if it feels big, and overwhelming, and impossible, I want you to understand that it's all normal, and to let yourself feel what's happening.

EMOTIONS ABOUT YOUR BODY

How do you *feel* about your body? The body you have always had? The body you have right now? The body you fear having? Write whatever comes up. You don't have to feel the emotions (yet). It's enough to just acknowledge them for now. That's always the first step.

HUMANS USE *LOTS* OF THINGS TO NUMB

Numbing-out is the opposite of feeling. The more you numb and avoid your emotions, the more you are adding to the pileup of emotions in the body that need to be processed and felt.

Humans use *lots* of things to numb themselves. Your phone, exercise, work, social media, alcohol, relationships, attention, sex—it's not always the thing itself, it's how you're *using* it. The very same activity can be used as a way to feel *more*, or to escape and numb your life, your body, and your emotions. Are you using certain activities or vices to be more in your body, or to be less in your body?

Let's use alcohol as an example. Alcohol is not *inherently* dysfunctional. There are lots of emotionally healthy people who drink alcohol as a way to relax or celebrate with their friends.

Problem drinking begins, though, when people use alcohol to numb or escape their lives, their pain, or their trauma. How much are they drinking to avoid their fear, boredom, rejection, and sadness? How much of a crutch does alcohol become? How much does it become a way to avoid dealing with your life?

Now, food is *not* the same as alcohol. We don't need alcohol to survive. We do need food, and lots of food, every day for the rest of our lives. You have the option to cut out alcohol if you want to reevaluate your relationship with it, but you *cannot do that with food.* We think we can treat our eating like a drug problem, but then we perpetuate the problem with our cure.

I'm sure you're wondering: *Okay, well, how can I make sure I don't ever numb out with food again?* You probably can't. You will numb out with food again as you begin to navigate learning to feel your emotions, and you don't *need* to be that perfectionistic about it to improve your relationship with food. Perfection is not required (or even possible). If you eat to comfort yourself, that's okay.

You might also be thinking: *Wait . . . but how do I even know if I'm emotionally eating or not?* Numbing out with food usually just manifests as faster-paced eating, less breathing, less awareness about what you are doing and what you need, tension in the body (but you're not paying attention to your body, so you probably don't even notice the tension), and some sort of desire to *shut down, escape, or run away from what's happening or what you're feeling.* The good news is that the switch from numbing to feeling just requires a shift in awareness and intention, and a breath.

That's it. A bowl of macaroni and cheese can be used to make you more embodied and connected (taking a breath and feeling what it feels like *while* you eat). You are allowed to be sad *and eat at the same time.* If you are willing to take an extra breath or two, before, during, and after your eating, you are moving in the right direction. So keep crying into your ice cream, friend, you're actually doing a really good job.

It's all about gaining awareness around when you're in your body and feeling, and when you're not. You can also start by changing your intention. Start *wanting* to feel. It takes practice and bravery to start feeling what's going on in your body, as opposed to numbing and escaping your body. But before you can do any of that, you have to decide, and want, to start feeling what you've been avoiding.

FEAR OF PAIN

Most of us have a fear of uncomfortable sensations. It usually goes something like this: we start to think of something uncomfortable, and we immediately try to shut it down because it's so *physically* uncomfortable. For instance, we'll have a flash memory of that time we said something embarrassing in a job interview. Just *beginning* to think about how *stupid* it was makes us want to crawl out of our skin. *Why did I have to say that third "How are you?" when I'd already asked and he'd already told me he was doing well and asked how I was???!* We begin to feel the prickly sensation of shame bubbling up, so instead of *feeling it* and reveling in *how stupid we must be*, we panic and do whatever we need to do to avoid feeling it. In fact, we never want to think or speak of it. Ever again. And this way of avoiding feelings becomes a perpetual, automatic habit.

But when you actually get down to it, the discomfort in your body is only sensation. It's constriction, buzzing, tingling, churning, hot, cold, spiky . . . and it's only our own associations with these sensations that's so horrifying to us. If we could have breathed through the memory of that third "How are you?," we would have given the prickly, hot shame and embarrassment a chance to *get out*. When we shove it down, that shame just stays, *waiting*. The big paradox is that the best way to get rid of any kind of emotional pain is to be willing to feel it.[58]

Another thing that happens when we avoid processing our emotions and experiences fully is that we start to fear what would actually happen to us if we *let* ourselves feel all the things we have been avoiding for decades.

We assume that feeling would be catastrophic. And in a way, we develop a fear of the fear. We think our emotions will swallow us whole. We fear we would just be consumed by the sadness or anger, that if we let ourselves go there and actually feel, we may never come back, that we may be sad or angry for the rest of our lives if it doesn't kill us first.

And then once we are dead, people will be so disappointed with us at our funeral. "Man, I never knew she was so *weak*. She cried and then she died. I guess she never deserved that promotion after all."

That's how intensely we avoid our emotions. And if you've experienced this kind of emotional repression, you know what I mean. But avoiding all our emotions and, by extension, avoiding our bodies, ends up compounding the situation, until we lose control altogether and melt down in panic, or burst out in anger, or break down in tears, and aren't even sure why.

Feeling won't destroy us, no matter what our deep, dark funeral fears whisper. Instead, if we can be willing to lean into the discomfort, and *increase* our tolerance for it, we'll be able to just feel it, and allow it, and it will process and pass.

We think that in order to master our emotions, we have to stuff them down and never feel or speak of them. We try to armor up, and then hope we can fool the world with our strength and steely exterior. *Then*, we figure, people will be very *impressed* with our lack of emotions and Botoxed facial expressions at our late but inevitable funeral.

What is way more uncomfortable than feeling the actual sensations themselves is that evasive, uncomfortable dance of shutting them down, repressing them, and avoiding them. But that's how emotions become stagnant, sitting there, waiting for an opportunity to make themselves known and felt again, normally in some sort of explosion.

Avoiding our emotions feels like constant dull, low-grade anxiety because we are constantly bumping into parts of ourselves that make us squirm and want to run away. We know what it feels like to live like this—it feels uncomfortable. But we are more scared of what would happen if we actually *allowed* ourselves to feel what we keep running away from. We figure the devil we know is safer than the devil we don't.

But the truth is, pushing our emotions down just compounds them and adds to the molten lava of emotions waiting to erupt later. When we actually *feel* them, emotions are able to process and pass.

Not only can it be uncomfortable to feel, but we are often *taught* not to feel. We have an unspoken social contract that public emotions are a sign of foolishness or weakness. *Oh, just get over it! Stop being so sensitive. Don't be weak.*

The more we think it's wrong to have emotions, the more we will take on different habits or vices to numb our feelings and emotions so we don't have to believe we are the kind of weak creatures who feel. Then we stuff down the emotions that are trying to bubble up, and it just perpetuates the cycle we are in. The emotions don't get processed, and they just stay stuck inside us, waiting and triggering panic, outbursts, or meltdowns.

Have you ever heard of someone crying from getting a massage or doing a really deep yoga stretch? That's because old unprocessed emotions stored in their body are being released. Lots of people who disapprove of "being emotional" will often find themselves experiencing an outpouring of emotions that they don't want and *don't approve of.* And when the emotions bubble up, it scares them so much that they attempt to stuff them down more, only to have them try to pop up again, further down the line.

This is also a lot like the binge/repent cycle. When we suppress something natural, it's going to keep trying to correct itself, and when our fear and disapproval makes us try to suppress it more, the cycle becomes even more extreme.

Feeling emotions is uncomfortable and scary in the beginning, especially after a lifetime of repressing them. There is a backup, like an overflowing clogged sink. Yes, yes. I am comparing emotions to a clogged sink. Do yourself a favor and let yourself finally feel them. Get out the metaphorical plunger, or whatever.

The paradox, of course, is that once you do the scary work of feeling what's in there, it becomes easier. It doesn't feel so hard or overwhelming anymore. People who actually feel emotions are the ones who are able to live their lives whole and present in their bodies, unafraid of being human.

WHAT WOULD HAPPEN IF I FELT IT ALL?

Before I ask you to start feeling the sensations in your body (which we will be doing soon!), first let's identify our *resistance* to feeling.

Write stream of consciousness on what you fear might happen if you allowed yourself to feel. What would happen to you? Who would you become? What would people think of you? What are your biggest fears, rational or irrational? There are no wrong answers; just write for five to ten minutes.

YOU CAN'T AVOID BEING HUMAN (SORRY)

Loooong before The Fuck It Diet, I was a total self-help junkie. And I assumed that I could *self-help* my way to better dieting. I was going to self-help away all of my emotional eating and transform into a skinny, skinny, enlightened body.

Obviously, the self-help diet method didn't work for me. Because diets don't work.

But it wasn't all wasted energy. All of those books I read primed me for The Fuck It Diet in other ways. And this important emotional part of The Fuck It Diet probably wouldn't even exist if I hadn't spent those years trying to self-help myself.

But back then, no matter what I learned, any sort of long-term mindfulness exercise where I tried to be more present, or watch my breath, or watch my thoughts float by . . . never clicked. I didn't *disagree* with it, I just . . . quickly forgot about it.

The only thing that really helped me was learning to come back into my body, which was taught in a really great way to me by my friend and teacher Alexis Saloutos. She has a master's in nutrition and has been on her own version of The Fuck It Diet for years. She sometimes calls herself a "nutritionist who doesn't work with food," and I originally studied with Alexis to learn more about how the body stores energy and emotions. Later on, we became good friends, texting constantly about the show *Outlander*, but that is a different story that I am willing to talk about for literally hours on end at another time.

Coming back into your body is all about the physical *location* of your awareness. It's as simple as just feeling what it feels like in your body—that is all that's needed to become present. This is so much more doable to me than anything else I'd ever learned from any self-help book, and also so much more profound, especially as someone who hated having to have a body that was any more than skin and bones.

We are meant to inhabit our bodies, but as humans with tons of unfelt and repressed emotions and a fear that we are too big or too much or too *whatever*, being in our bodies is really uncomfortable, so we live most of our lives preferring not to feel.

Instead of being present in our bodies and feeling what's there, we hang out up in our minds, overthinking everything to death and feeling as little as possible. We have this hope that we can transcend our bodies and fix all of our problems with our minds—but that just doesn't work. Our attempts to heal physical and emotional wounds simply by thinking about them won't do anything. You have to be *in* the body to heal the body. You have to come back down.

Basically, we are all humans, running around, not wanting to be humans. We don't want to have to *feel* what it feels like to have a body or have emotions. If you think about it, you can look at the fear of food, or the fear of our bodies, as inherently *anti-life*. Literally . . . if you do not eat, you will eventually wither away and die. This is especially true for more extreme eating disorders, like anorexia, but it's also an underlying dynamic for the chronic dieter who is constantly trying to become smaller and more "acceptable." The idea of wanting to weigh less and less and less can be a subconscious desire to get *closer* to not existing, not feeling, not having to *deal* with the hard stuff.

Don't get me wrong: *I get it.* Life is fucking hard. Having a body is *hard*. Having emotions *is hard*. And it makes sense that we think dieting could be the thing to save us from the pain of existing. Not only does losing weight make us more praised and accepted in our culture, but remember, dieting is also a really good way to numb and distract us from our emotions and from what's really going on.

Many, many people with any sort of disordered eating have a *lot* of difficulty feeling what it feels like in their bodies. Their awareness is completely up in their head. And a big game changer for *me* was the idea of bringing my awareness "back down" into my body, and being fully willing to feel and be a human. Being willing to be here and feel what it feels like to have a body and take up space, as opposed to withering away into vapor. Fixating on weight loss is an attempt to leave the body—to shrink and take up as little space as possible in this world. It also helps us to *feel* as little as possible. It's a big glaring manifestation of not really wanting to be here and not wanting to *deal*.

It can be really uncomfortable and difficult in the beginning to come back down, mostly because of all of the emotions and discomfort that will bubble up. These are emotions that wanted to be processed and felt before—and have been waiting for you—and feeling them is the only sane way forward.

When you eat, you are actually bringing "the earth" into your body—tying you to the planet and keeping you alive. It's bringing *weight* to your physical existence. The act of eating and coming back into your body is asking you to accept being human. It is asking us to integrate with the most uncomfortable, messy, earthly, painful, and base parts of our existence.

Being in your body is also where you are *most* alive and vibrant and powerful and connected. And almost all of us are resisting it.

"Coming back down" is asking you to actually feel and inhabit the body you have believed was too big and too much and too ugly. It is asking you to actually take up space, whatever space your body naturally wants to take up, as opposed to shrinking up and away, and wanting to become a little skin-and-bones fairy who doesn't need to deal with earthly woes like food and emotions and fat. We need to become willing to feel what it feels like to have a body. And this applies to anyone, no matter what their natural body size is. In the simplest terms: not wanting to be in your body is not wanting to be fully alive and fully human, inhabiting the earth.

The origins of modern dieting also started from a religious and body-fearing worldview. Sylvester Graham, who invented the Graham cracker in 1829, was a Presbyterian minister who believed that fiber and whole grains would lower sex drive. He pushed a vegetarian diet, and believed that "delicious food" (his words) and meat and coffee would lead us into sin.

In the late 1800s, John Harvey Kellogg expanded upon Graham's teachings. Yes, this is the original creator of Kellogg's flaked grain cereal, *and* what we know of today as granola (though his granola was meant to be tasteless and sugarless, and lower libido). Kellogg was a very religious doctor who believed that bland food, fiber, and a "plain and healthy diet" would curb libido and masturbation.[59] (The irony of course is that "healthy eating" to curb sex drive is contradictory in and of itself, because lower sex drive means something is off—like maybe you are starving or dying and your body has shut down procreation.) He *also* prescribed genital mutilation on both men and women to curb sexual appetite, he was in favor of racial segregation and eugenics, *and* he practiced complete abstinence—he never consummated his marriage with his wife, and they adopted all of their children. Cool guy.

So John Harvey and his brother went into business to sell an antimasturbation cereal: corn flakes. True story. His brother Will cared less about sexual purity and more about business, and wanted to put sugar in the cereal recipe so it would actually be palatable. But John Harvey Kellogg was *very against this idea because sugar would increase sex drive*. It started a feud they never recovered from and, lucky for us, Will ultimately got the Kellogg Company.[60] Long live Frosted Flakes!

This connection between fiber, "healthy" eating, food purity, spiritual purity, and deep fear of our bodies and our sexuality is profound and *should not be ignored*. This was the very beginning of our modern dieting. Don't discount how the centuries of fearmongering about "the sins of the flesh" have gotten into our collective psyche. We are *still* assigning morality to certain foods and certain ways of eating today. We are still afraid of what having an appetite *means* about us. We are still assigning morality to certain *body types*. And we are still afraid to be too hungry, too decadent, or have too many desires.

Seeing your weight and your body as *an issue* will not help you. Instead, accepting, feeling, and inhabiting your body in whatever state or shape you currently are in is really, really important. Your body is constantly asking for you to come back down and come back home, so let's just give it what it wants already.

EATING TO HELP YOU COME BACK DOWN

Can you begin to see eating as physically bringing you back into your body and grounding you? Can you start to see that feeling and taking up space in your body is an important part of coming back down and being willing to be fully human?

Use this concept to help expand your awareness of your energy and what it feels like to be grounded and inhabiting your body—and what it feels like to have a better, fuller relationship with eating.

WE'RE ALL STUCK IN FIGHT-OR-FLIGHT MODE

The good news for you is that all of this woo-woo feeling stuff is also biologically sound, and all has to do with the fight-or-flight mode I mentioned earlier. In his book *Waking the Tiger*,[61] Peter A. Levine, PhD, explains that in *any* shocking or potentially life-threatening situation, the primitive part of our brains will unconsciously mobilize adrenaline and energy in the nervous system to prep us to fight or flee. But often humans don't let this instinctual fight-or-flight process naturally complete itself and discharge the energy we have just built up. The result is that the body remains in high alert long after it needs to be—sometimes for our whole lives.

And the only way to heal and get out of this state is by feeling the old, stuck sensations . . . by getting into our bodies, breathing, and feeling.

We keep messing with our biological processes and getting stuck in survival mode. We keep accidentally interfering with, and overriding, how our body is wired. We keep judging the mechanisms that are in place to heal us.

So now there are *two* survival states wreaking havoc on our quality of life: one is the famine survival mode, and one is this fight-or-flight mode. Both will have us running on adrenaline for way longer than is ideal, which will wreak havoc on our bodies, override our brains, and totally deplete us for years on end. We need to activate the rest-and-digest mode, but we are stuck.

This high-alert fight-or-flight state is what we have come to casually refer to as trauma. The nervous system *believes* the threat is still present, even when we are perfectly safe. Some people explain trauma as frozen energy in the nervous system and body that needs to be somatically felt in the body in order to "thaw."

This is not to diminish the experience of people who have gone through truly terrible or abusive experiences, or to say that "everyone is equally traumatized." If you are experiencing clear and obvious trauma or PTSD that is overwhelming you, please seek the professional help you deserve to guide you through healing. And if it ever feels like "too much" during any of the exercises I give in this book, just stop and go at your own pace, and seek support from mental health professionals.

For the rest of you who aren't sure if this fight-or-flight thing applies to you? On a biological level, it probably does. *Very* few of us get through life completely untraumatized, and you probably have pent-up stress in the nervous system from any number of unprocessed experiences, especially if you tend to resist being in your body, or if the idea of coming back into your body is uncomfortable to you, or you feel easily stressed or overwhelmed.

Levine explains that trauma isn't just the result of a truly life-threatening situation. It can come from something way less life-threatening that the nervous system still responds to as though it was life-threatening. Meaning, your body could very well be experiencing varying levels of trauma from something that you logically know was never life-threatening at all—even surgery, dental work, an innocuous experience from childhood (like thinking you've been separated from your parent in a store), or a close-call car accident—as well as a host of emotional and social traumas, like heartbreak or social shaming.

Humans experience trauma at such high rates,[62] compared to animals, because we are rarely in our bodies. We are thinking instead of feeling. And because of that, we don't let the biological process of fight-or-flight complete itself. On the other hand, wild animals just roll with their fight-or-flight response by being fully in their bodies, having no option to jump out of their bodies and into a reasoning mind. They are able to recover quickly after shock and "discharge" or "process" the energy, often by involuntary shaking. They allow fight-or-flight to do its thing, and therefore they rarely experience lingering effects of trauma. But we humans accidentally halt the process. Instead of feeling the intense sensation in our bodies and properly allowing it to "discharge," we let our brains get in the way. We are afraid of the intensity of the sensation, so we stop feeling and we start rationalizing. We don't let the survival response do its thing, and it wreaks havoc on our bodies and minds long-term.

The way out is through feeling what is in your body. Actually staying with the raw sensations that are happening in your body. Notice what they feel like, and let them move.

This next tool will help you stay with—and process—emotions and sensations in your body, and the heart of this tool is simply breathing and feeling. It is going to sound frustratingly simple, maybe even *too* simple, but it's not. Remember, this is a building block toward a different way of dealing with your unresolved *stuff*—and it will help you get used to surrendering and feeling, instead of running away.

» TOOL #3: BREATHE AND FEEL

Set a timer for five minutes and lie down. Your only task for these five minutes is to feel the most intense sensation in your body, and breathe into it. Ask yourself, *What is the first feeling that pops out to me? What does it feel like?* Then breathe. Don't try to change the sensation or make it go away—let it be. Be curious, and see what you can learn about how that particular sensation feels: Is it hot? Cold? Is it moving? Is it pulsing, prickly, churning, sharp? How big is it? What color is it? How dense is it? How intense is it on a scale of 1 to 10? What do you notice the most?

If this sounds like a meditation to you—you're right! It *is* a form of meditation. A short, focused one.

IF YOU'RE HAVING TROUBLE FEELING SENSATIONS, just start by feeling your body in space. What do you feel touching your skin? What does your skin feel like? Now what does it feel like beneath your skin? What *do* you feel? Then you can ask again, *What is the most intense sensation in my body right now?* Then breathe into it. Use breathing to feel *more*, not less.

BUT THIS IS SUPER UNCOMFORTABLE . . . Yes! Great! You're doing it right. Lean into sensations that are uncomfortable and see what they really feel like, and what they're really doing, on a very basic sensation level. And it's only five minutes long, so the discomfort won't last forever.

IF YOU FORGET WHAT YOU ARE DOING, and why you are lying there, and start thinking about the fact that you should have put your clothes in the dryer, that's fine. That's life. All you need to do is gently redirect your thoughts back into the body by breathing and then feeling, asking, *What is the most intense sensation in my body?* Just work on the habit of feeling. That's all you can do.

And *now* when people ask you, "Do you meditate, bruh?," you can say, "Yeah, bruh, I meditate. *And* I lie down. I'm doing *great*."

You can also try doing this kind of deliberate breathing and feeling in your everyday life—walking around, or reading emails, or whenever you inevitably come up against experiences that are uncomfortable—but one step at a time, bruh. One step at a time.

THE MYTH OF STRESS AND BREATHING

There is a myth about the way breathing reduces stress. We think that becoming calm is just a couple deep breaths away. "Just calm down! Take some deep breaths!" But that doesn't always work for everyone. In fact, sometimes breathing can initially cause people *more* anxiety, and that's because when we breathe full, deep breaths, our awareness is brought back into the body, where we are forced to feel *more.* Breathing activates old/stagnant energy and emotion (or more fully brings you in contact with your *current* stress), so you may actually feel more uncomfortable before you feel better.

But we *want* to get into the habit of breathing into discomfort, because that is how to stay with it and process it in real time. If we were always in the habit of breathing and feeling, we would be ready to process our feelings when they first arose instead of avoiding them and storing them for later panics or outbursts. But since we don't feel in real time, that energy becomes stagnant and stored for later.

The most concrete way to point to energetic and emotional stagnation is in your muscles. We all hold unprocessed emotions in our tissues and muscles. The muscles tense up and hold emotion and stagnant potential energy until we intentionally circulate and feel it through massage or stretching or breathing or any other kind of deliberate feeling or energy work (like the breathe-and-feel tool).

Emotions are also stored in organs and other tissues of the body.[63] Neuropharmacologist Candace Pert, PhD, believed that the body *is* the subconscious mind, and has seen that our glands and organs have peptide receptors that can access and store emotional information. She stated, "The real true emotions that need to be expressed are in the body, trying to move up and be expressed and thereby integrated, made whole, and healed."

Eastern medicine also correlates different organ systems with different unresolved emotions. For instance, liver stagnation is often associated with unresolved anger in the body. I remember one time I was *furious* lying on the acupuncture table, and I told my acupuncturist when she came back into the room that I was *super* frustrated and uncomfortable and pissed off. She said, "I'm not surprised. I'm working on your liver."

The breathe-and-feel tool was adapted as the simplest way to process my own stagnant emotions, and to guide TFID students to process theirs. There are many nuanced healing methods to balance the body, or guide people into their bodies to help move what is stuck: massage, acupuncture, acupressure, and yoga, or whatever obscure methods strike your fancy (reiki, tapping, Rolfing, etc.).

You can seek out *any* kind of body work, energy work, or movement that helps you get into your body. You can try any and all methods—but you can also keep it super simple and just breathe and feel. They are all just different ways to help us get into our bodies and feel and process. Understanding that will give you a leg up.

THINK YOU CAN HATE YOURSELF INTO IMPROVEMENT?

When I was four, I was dancing around the family room and fell hard and hit the top of my cheekbone on the corner of the TV stand. I got a big black eye and a scar that I have to this day. I cried a lot, because I was *very* dramatic, and my parents bandaged me up and debated whether to take me to get stitches (they didn't, hence the scar).

"Ooh, this is actually a really deep gash. Should we take her to the emergency room?"

"First use a butterfly Band-Aid to keep the cut closed."

Butterfly Band-Aid?!? My parents hid secret butterfly Band-Aids that I didn't know about? Why would they have kept that from me? The promise of a butterfly on my face paused my hysterics for a moment, until I saw them open up this stupid white bandage that looked nothing like a butterfly. Through tears I said, "That doesn't look like a butterfly . . ."

"No, this is a butterfly *Band-Aid.* It is better than normal Band-Aids at keeping skin closed when you have a deep cut."

I think I started crying over my broken face again. And at some point I started repeating over and over, "Ughhhhh, I hate myself I hate myself I hate myself I hate myself."

My dad said, "Caroline, why are you saying that??" My parents never scolded me for getting hurt, there was no logical reason I'd be so hard on myself, but I remember feeling embarrassed and angry. I shouldn't have been so *stupid*, like a little *child.*

Even at four years old, I wanted my parents to think I was a grown-up. And this injury was just hard proof that I wasn't. This was proof that I was actually just a little kid who was stupid enough to twirl around the TV case—even though they warned me that I was twirling too close and might fall—and I did it *anyway. I hate myself. Why am I so stupid?*

If I hated myself openly, then everyone would know that at least *I knew I should do better*, and it would teach me to stop being such a stupid little kid who doesn't have ultimate control over my twirling . . . *and maybe I shouldn't even twirlllll. SOBBBB.*

"I shouldn't have fallennnnnnnnnn."

"Caroline, this wasn't your fault, people fall all the time."

But I'm supposed to be better than those other people who are stupid and fall when they twirl.

I honestly don't know what past life horrors taught me to be such a weird little perfectionist, but it took me a very long time to unlearn.

Does this sound familiar to you? Because this kind of self-blame is what dieters do. *I'm supposed to be better than all those other people who can't lose weight! And if I'm not, I'm supposed to be very angry with myself for my laziness and my failure and my twirling.*

We think that hating and berating ourselves will be the thing that keeps us improving. And we think that if we start being *nice* and *understanding* toward ourselves, then we will just start accepting laziness and ugliness . . . and *then* we might actually even be *happy* when we are absolutely NOT allowed to be happy yet! We don't look good enough to be happy yet! We haven't bled enough to earn being proud of ourselves. Happiness isn't allowed while big companies are paying billions of dollars in advertising to constantly remind us now *not-happy* we should be!!!

And when we see someone actually happy at a higher weight, it can feel so foreign to us—because we are consumed with our own fears of weight—that we decide they must actually be unhappy. We think that we need to hate ourselves into more self-improvement. We think we need to shame ourselves into becoming more responsible and beautiful.

It seemed easier and safer to hate myself *first* for failing on my diets and gaining weight—before other people had a chance to.

Before they had the chance to be disgusted with me, I'd be disgusted with myself. And if I was disgusted with myself enough, then I'd be inspired to turn my life around once and for all and become super controlled: go to the gym every day, lose my taste for carbs, and become extremely thin and beautiful. And beautiful equals happy, duh. It was simple. Hate would lead to happiness.

I also tried to *trick* myself into not wanting to eat by using really absurd and strange mental gymnastics and fantasies. For example, one time, in college, I wondered if I could successfully diet by pretending I was a vampire—and that all desserts were my true love that I couldn't eat. Thank you, *Twilight*. I also wondered how Harry, Ron, and Hermione could think so little about food, and I figured I just needed some sort of big magical wizarding world war to concern myself with, because, clearly, that's better than wanting to eat.

I don't think I need to tell you how much this didn't work. I was never able to keep up the "self-improvement" when the motivating factor was disgust. You cannot hate yourself into happiness. That's just not how happiness works.

Sure, you'll feel that fleeting euphoria when you get a raise or get a compliment for losing five pounds. You've "won." Our brains *love* winning. But we have been confusing that high with real happiness. Without a steady underlying acceptance, euphoria leads to a crash every time. And then the highs become an addiction itself. We start chasing quick-fix feelings of approval or winning, only to wonder why we are still so insecure and miserable.

Chasing these highs is the real addiction you should be worried about, not carbs.

WHO CAN WE BLAME????

When you begin to go through The Fuck It Diet, it is normal to get angry and want to figure out who to blame. You've wasted years carrying pointless, painful, debilitating self-doubt—because you were taught to. You used to direct anger toward yourself to try and inspire more willpower. And it's time to use that energy in a different way.

Getting angry has benefits. It can inspire you to stand up for yourself, empower you to have a differing opinion from other people and to create strong and healthy boundaries. Using that energy of anger and rebellion can be a really healing part of this journey, and an amazing way to reclaim your power and feelings of worth.

And remember, if you feel anger, you can't just ignore it. Especially if it is old and repressed, you're going to need to experience it so it can process and pass, otherwise it'll rule you from backstage.

Many of you will have to figure out new ways of interacting with the people you love. Friends, family members, partners, or previous diet buddies may start to make you really frustrated and overwhelmed, because they are still imprisoned by the fixation on weight and food—and still want you to be too.

It'll be tempting to make those people the enemy. But as frustrating as their comments and judgments may be, they have simply believed what they have been told about health and beauty and responsibility—just like you did. They believe that being thin is important and safe and healthy, and that perfect eating is important and healing, and maybe that they should constantly be telling everyone else about it, too, for everyone else's own good.

So many of the people I have worked with have a lot of anger toward their families, and often their mothers: many people's original diet partners. One of my student's mothers is eighty and still considers sugar evil, and still comments on every food her fifty-five-year-old daughter puts in her mouth.

So many people are obsessed or overly fixated on food, purity, weight, exercise, and looks. And yes, if and when these people are your parents, this can be the cause of a lot of disordered eating in children, lots of self-judgment, and lots of pain. Surprise, surprise! Our parents' neuroses and beliefs shape and affect our own before we ever have a chance to *un*learn.

The most powerful way to move forward is to accept that they tried but they don't know what they were doing, and they don't know what they are talking about. At the root, most people have good intentions. Even if they come off as tone-deaf assholes in the process, they just want to help you. They think they actually *are* helping you, and that you are just too stubborn to listen. And, hey, you may have even been like this too—maybe not that long ago.

So try to understand that your mom, or grandmom, or dance teacher, or doctor, or your friend who swears that she has actually found a diet that works . . . they are all just trying the best they can. The sooner you can understand that and realize that *you* are now freer, no matter what they think, and understand that your worth has nothing to do with weight, or their beliefs about weight, the easier this all will be. And as frustrating as people may continue to be, seeing us all as victims of a bigger paradigm will make it easier to exist alongside them in a world that values thinness above (almost) all.

You are allowed and encouraged to take space from any and all people who can't stop talking about diets. And you are allowed to *tell* people to stop talking about diets and weight around you.

Relating to people who fundamentally don't agree with you, or who are blinded by what they have been taught to believe about bodies, is one of the most frustrating parts of The Fuck It Diet—and, in all honesty, one of the most frustrating parts of life. It can be disheartening and even heartbreaking. But at the end of the day, all you can do is work on yourself, and get clear on what you believe. I will elaborate on more specific ways to do this later in the mental part, when we talk more about beliefs.

Using anger as fuel to protect yourself, stand your ground, take up space, and make up for lost time will help to reverse the silencing and oppression and self-hatred. Anger will be helpful in the beginning. But eventually what may be *more* sustainable is the understanding that most people are oblivious to what is going on. Most people think they can help you be happier and healthier. Most people don't really know that they are spewing random dogma and that they, themselves, are victims.

Then there are many other people who develop obsessions with eating and weight, not because of their family but because of the media and culture at large being *obsessed with thinness and fitness.* I recommend getting angry at the culture that convinced you that you weren't enough, and realizing we are part of the culture, and figuring out what little ways we can help change it.

You can fight to change beauty standards, you can explain the new information you are learning, and you can also try and appeal to people's humanity and ask them to respect yours—and I recommend you do. But you can do that by *not* living in anger, instead *using* anger to rebel, processing your emotions, and choosing to be optimistic. I am a big fan of rebellion plus optimism. Believe that the new way you look at the world, and food, and bodies, could become the next generation's way of looking at the world.

On the flip side, healing your own "stuff" with food and weight can also have a healing effect on relationships. One of my students shared with me, "I am noticing that The Fuck It Diet has *actually* made me CLOSER to friends that are still dieting. I feel compassionate and also neutral and separate from them and their dieting, and I am now naturally interested in exploring topics other than diets, so we've been able to connect in a deeper way. This is something I didn't expect, and it's so cool to experience!" TFID for the win.

Most of all, remember that you are allowed to be happy, even when people don't understand why you are. Try something like this on for size: "You can think that I am not dedicated enough to fitness, but I don't give a fuck what you think. I hope you're genuinely fulfilled and happy with your intermittent fasting."

Remember: Fuck it.

WRITE A LETTER TO YOUNGER YOU

Write a letter to a younger version of yourself. You get to pick the age, but make it a time when you needed someone to understand you and lift you up. Write to younger you, knowing what you know now. A nice age to choose could be right when you started dieting, or maybe when you were in the thick of your diet misery. What does this younger you need to hear?

Then, if you feel up to it, travel back in time and write a letter from your younger self back to yourself now.

Then get in your Tardis DeLorean time machine and go find out if it was Adam and Eve or the Big Bang and evolution, then @ me on Twitter and let me know so we can spread the word. Do not step on any butterflies. This is called the butterfly effect for two reasons. Okay, go write.

SURRENDER TO THE MESSINESS

Feeling is messy. Emotions are messy. So is this whole Fuck It Diet. Nothing is linear or straightforward. It's all messy, and it's okay. This is the way it's meant to be. It's teaching you to surrender to the messiness.

One student realized, "My orthorexia and fixation on purity

was actually an obsession with being 'special'—being different, unique. Being obsessed with food purity made me feel elite. I was too good for restaurants, too good for my parents' cooking. Letting go of this control has lifted the weight of the world off my shoulders."

We've hoped that we could use dieting or weight loss to help us feel special or untouchable, which is just another way we try to feel worthy of love or adoration. The alternative to allowing and feeling imperfection is shutting down and deciding we can never address that we failed—and never feel the emotions that come along with failure.

Humans are always going to be messy and imperfect. I don't really see how we can expect to heal without surrendering to that and allowing ourselves to go through the messiness of healing and learning phases. Healing actually comes when we are finally willing to admit that and *feel* what comes along with that imperfection—again, it's about *feeling what is there*, instead of pretending it doesn't exist. Right now, in the middle of this messiness, is exactly where you're supposed to be.

I know you've probably heard this plenty of times, and rolled your eyes or said, "Sure sure sure, yeah yeah. I'm exactly where I should be right now. Blah blah. Now let me lose some weight." But the cure for shame over our bodies is not losing weight, or becoming more perfect, or making more money, or shutting down all acknowledgment of how we feel, or controlling every little thing. Perfectionism is just a temporary shield that will never help what you're feeling on the inside. All you'll be doing is putting up more walls while you continue to crumble inside, petrified that the world will find out what your arms really look like. Or learning that *GASSSP* you don't actually have it all together. It's just more walls, separating you further and further from the real world.

Actually *feeling* the emotions that come along with being imperfect, embarrassed, wrong, rejected, or making mistakes is the key. That's what everyone is talking about when they mention "vulnerability": feeling and letting it in and knowing that it will not destroy us. Feeling allows us to process and become stronger, more whole and integrated people. Learning to feel will not only give you new ways to process what we used to numb, but it will help empower you to embrace the messiness. No perfection needed.

THE MENTAL PART

The mental part of this process is about working through all of the unhelpful and painful things we have *learned* about weight and our bodies. These beliefs are often what make The Fuck It Diet so hard—even if by this point you are feeling way better with your eating, you'll still want to clear out a lifetime of food rules, food guilt, and body beliefs, because if they go unchecked, they can wreak havoc from the background, keeping you feeling stuck and afraid. This is the phase where we become aware of, and heal, the "mental restriction" that is still lingering. We are going to begin to unravel and identify our limiting beliefs.

Many people are tempted to try The Fuck It Diet while "watching their weight." Beware . . . if you are not willing to face your fear of gaining weight, you will end up where you started. I know it's really hard. If it were easy, weight wouldn't be the huge drama that it continues to be in our culture and personal lives, and you wouldn't need this book. Our beliefs about beauty, weight, and worth are at the heart of our complicated and disordered relationship with food. Being willing to unlearn what you have learned, and change your beliefs about weight, will make all the difference in the end.

You cannot learn to eat normally if you are unwilling to understand how you got here.

UNTANGLING THE KNOT

I like to describe our subconscious as a big complicated tangled knot of unhelpful beliefs. In the beginning, everything is jumbled and confusing and stressful, and it's so easy to get triggered and spin into panic because all of the wires are crossed. We are not even fully sure what's at the root of our panic, but it's connected to a million little thoughts, fears, and beliefs that are continually triggering us.

Untangling that knot is slow, patient work. You can't do it all at once. You can't just pull one string and unravel the whole thing, because this knot is way more complex than that. Each mini knot that is part of the bigger knot is comparable to an unhelpful belief that is contributing to the larger anxiety-inducing mega-knot. Sometimes pulling one thread will even make a different knot tighter. Go slowly and deliberately and compassionately, thread by thread. And you have to be nice to the knot, because the knot is also sentient.

In the beginning, when the knot is extremely jumbled and tangled, it's hard to see how the threads are connected. Because it's a mess. It's overwhelming. But the more you unravel it, the easier it is to see, and the clearer the next step becomes.

There is a complicated web of anxiety, created and perpetuated by smaller knots—so work on the smaller knots first. Identify where the anxiety is stemming from, then slowly and patiently unravel it. The less tangled your whole mind is, the more clarity you have, and the easier it is to see *where* the stress is coming from, and *why* you are feeling it. You will be untangling big and little knots for the rest of your life, but the process becomes easier and clearer as the whole web becomes less tangled.

This metaphor is also meant to illustrate that yes, you're a big freakin' knot that you'll need to untangle, but you are not broken. Snipping away at the tiny knots isn't gonna help either. I don't know what snipping is symbolic of, maybe a lobotomy. So, let's get #nolobotomies trending, and begin a slow untangling of all your unhelpful beliefs.

» TOOL #4: THE BRAIN DUMP

We can't untangle anything if we can't even see the knot, so this tool is the first step to help us do that. We can't heal if we don't even know what the problem is! The brain-dump writing exercise will help you see what is *really* going on underneath the surface. It will give you way more awareness. And awareness is always the first step.

The way to do the brain dump is to write for twenty minutes straight, stream of consciousness—everything you think, feel, and worry about in the moment. Get it out of your messy brain, and onto the page. That's it. It's that simple.

I used to be incredibly anti-journaling. I thought it was duh-ummmmmb. *Two syllables worth of dumb.* That's how dumb and pointless I thought it was. Plus, I was a *writer*, so if I was going to *write*, it was going to be something *brilliant* that I could share or save forever. The idea of journaling as a therapeutic practice just annoyed me. I also thought there was no way it could actually help or *do* anything. And now I must fully admit how wrong I was, because brain-dump journaling changed my life. It guided me through The Fuck It Diet and all the realizations and feelings I had as I went along. It is probably one of the things that helped me *the most*. Do not underestimate this practice.

Taking twenty minutes to do a brain dump is a way to get clarity on what is going on in your mind, heart, and subconscious. Nothing is too mundane. Nothing is too petty. Just write whatever comes to mind. No editing. No stopping. No judging what you write. Write down what is taking up space in your mind, and what you're worried about today. Don't judge it as it's happening. Don't censor yourself. Get it out of your mind and onto the page where you can actually see what is in there.

Our thoughts will always be crazy and neurotic and worried; it's just the way the brain is wired. Our worried, petty thoughts will never stop happening, so instead of being consumed by them, we can just become more aware of them, and know that it's just a normal, mad mind, yammering away. *Hello brain, I see you, you little bitch. Thanks for trying to ruin my day.* Lean into the madness.

If you want to truly heal, you have to stop ignoring what is going on in your mind, heart, and body. This is a way to take the mental chaos, look it in the eye, and make sense of it.

You can do a brain dump whenever you are stressed, overwhelmed, emotional, confused, when you start panicking about the fit of your pants again, and any time you need clarity or guidance.

When I use the brain dump, I range from talking about all my fears and stressors, to writing my to-do list, to brainstorming ideas of what to say or how to respond to an email or how to organize an online course I want to create, to fantasizing about my future, to making jokes to myself, to asking for help/guidance, to writing down ideas for tweets . . . you name it. It is a brain dump. You can't fuck this up. Just write whatever. During the first six months of The Fuck It Diet, almost all I wrote about was food and weight. And slowly, as my focus changed, the content of my brain dump changed too.

Keep in mind that this is not a diary. You are not passing these journals down to your children and grandchildren. In order for you to feel comfortable writing about what is *really* going on and how you *really* feel, these are fully disposable. Throw them out right away if you want. Shred them. Or you can wait and fill up a brain dump journal and then throw it out. Do not be precious about this writing. Full sentences are optional. Saying something just the right way is optional. You can read it back, or never read it again. The work is the writing alone. By writing everything that you are thinking in your mind, in real time, you are doing a meditation on the page. You are using this stream-of-consciousness writing to *watch* your thoughts and to notice patterns and make a little space in your mind by getting out the jumble and looking at it on the page.

Adapt your brain dumps to work for you. Do them once a day, do it five times a day. Have a ritual or use it as needed. I recommend twenty minutes a day, but you can do it for ten minutes or an hour. All we are trying to do is get some more clarity and ease by slowly but surely *examining* what is making up this knot. Just do it. It will help.

THE POWER OF OUR BELIEFS

Once upon a time, there was a regular old milkshake that was part of a study. This study was measuring the ghrelin levels in the people consuming the milkshake. Remember, ghrelin is the "hunger hormone," and when ghrelin rises, it signals to your body that it is time to eat—so high ghrelin levels equal hunger. Low ghrelin equals no hunger. And when this hunger hormone rises, it *slows down* the metabolism "just in case you might not find that food," according to Alia Crum, the clinical psychologist running the study.[64]

This milkshake was given to two different groups to drink. The first group was given the milkshake in a cup labeled "Sensishake: fat-free, guilt-free—140 calories." That group thought they were drinking a fat-free, low-calorie "health" shake. The other group was given the *same*, normal milkshake in a cup labeled "Indulgence: Decadence you deserve—640 calories."

In reality, the milkshake was somewhere in the middle: 380 calories. And before you get all prickly over me talking about calories, I promise it's worth it.

When the participants drank the milkshake that they *thought* was 640 calories, their ghrelin levels dropped. They weren't hungry anymore.

But when the participants drank the milkshake that they thought was only 140 calories . . . their ghrelin levels *didn't drop*. The hormone levels stayed high, and they stayed hungry, and their metabolism stayed slow, *even though they were drinking the same milkshake that the other people were.*

Do you hear that? The very same milkshake had completely different effects on the two groups, based on what they *believed* they were consuming.

"Our beliefs matter in virtually every domain, in everything we do," Crum said. "I don't think we've given enough credit to the role of our beliefs in determining our physiology, our reality."

That means that what you *believe* about what you're eating has more power over your physiological response to food than the actual caloric and nutritional value of the food. Your body is heavily affected by what your *mind* is doing and thinking. And *that's* why just *considering* going on a diet can spark a binge.

Thoughts about "needing" to lose weight and cut back on food can make the body feel the effects of deprivation before you even start physically restricting, which will automatically affect your hunger and satiation hormones. Just *considering* going on a diet can signal your body to keep your ghrelin levels high, making you physically hungrier than you would be if you were planning on eating whatever you wanted. The body remembers those diets you went on. The body will not be fuckin' fooled.

This phenomenon is the *mental restriction* I have been talking about. And it is why your negative and fearful beliefs about food and your body need to be addressed, in order to truly get out of this dysfunctional cycle.

If the way we think about the food we are eating has a physiological response in the body, I like to encourage people to take this concept and run with it . . . because most people have been believing that everything they eat is killing them. They think they shouldn't be eating something *while* they are eating it. How helpful do you think that is? What kind of physiological response does that create in the body? A shitty, stressful one. That's what.

YOUR BELIEFS BECOME CONFIRMATION BIAS

Not only do our beliefs directly affect our body, but beliefs shape the way we see the world around us. This is a psychological phenomenon called "confirmation bias," where we filter evidence and interpret everything as a confirmation of our *existing* beliefs or theories.

This is how people get into untrue conspiracy theories, where everything they see supports the theory they already have, even if they have to twist it to make it fit. That's also why and how our country and our world are so divided; that's how two sides believe what seems like two completely different sets of "facts."

Many of our beliefs are subconscious. We aren't fully aware of them without deliberate inquiry, but even from the background they can shape the way we interpret and interact with the world. We experience what we believe.

The beliefs that are stressful or negative are often referred to as *limiting* beliefs, because they are literally limiting you and your life experience. All of your beliefs about money, love, life, happiness, and health are affecting you practically, subconsciously, and energetically, *and* your beliefs are being mirrored back to you through the way you interpret your reality. We pay attention to the things that confirm our beliefs and ignore the things that don't.

This concept of limiting beliefs applies to everything, including your relationship to food and weight. And your beliefs about food and weight are the things making this process so damn hard.

On The Fuck It Diet, beliefs can sneak up on you months down the line. My student recently shared, "I was going along, and feeling so much better about food, and then I felt the old anxiety come up again, like I was waiting for the other shoe to drop. I realized I was treating this like other short-term diet success I'd had in the past. I believed 'IT WON'T LAST.' When I examined these beliefs more closely, I realized I don't *have* to wait for the other shoe to drop. This isn't about 'good' or 'bad.' Remembering I'm never going back into the famine has chilled me out."

The best way I know to work through mental restriction is gaining awareness around your beliefs, and the best way I know to help you gain awareness around your beliefs is *writing* (the brain dump). If you're able to pause and try to identify lurking beliefs in the moment without pen and paper, more power to you. But I've found that writing it out not only helps you find your limiting beliefs but also helps you remember them and refer back to them.

When beliefs lurk in the dark corners of our subconscious, we don't even realize they are controlling us. So shine a light on them. Having awareness about what's ruling you is key. Look them square in the eye. Say: *I see you, you little fucker.*

FIND THOSE LITTLE FUCKERS

Sit down and write out a list of all the negative, limiting beliefs about food and weight you can think of. The list will probably be really, really long. Remember, the first essential step is awareness.

Remember those food rules you wrote out back in the physical section? Most of those are limiting beliefs. Look back through that list, and identify any that still feel particularly true or stressful to let go of.

How do you know if it is a limiting belief? Does it cause stress? Yes? Then it is negative, or "limiting."

Begin to consider that most of your stressful beliefs are untrue. It is the first step.

MENTAL RESTRICTION AND BINGEING

If you are still bingeing, you are still restricting in some way. And if you can't figure out where you are still restricting, it is most likely coming from *mental* restriction and guilt around eating. In my experience, doing work around your negative, limiting beliefs is the easiest way to slowly unravel and heal lingering mental restriction.

Bingeing is *not* happening because you are a totally out-of-control food addict but instead because you either aren't eating enough yet or still don't trust that you're actually *allowed* to eat, and that lack of trust will continue to get in your way.

This is true even if you're not *quite* bingeing, but just constantly low-grade overfilling—the answer is the same. Eating in a way that makes you feel out of control is either from restriction, or a resistance to being in your body and feeling, and often both at the same time.

All the food rules, food guilt, and diet beliefs that you have not quite gotten rid of yet are still affecting your eating and your appetite. The body doesn't want you to go on a diet or be starved. The body wants you to eat. So if you are bingeing, your job is to figure out where you are still restricting.

If you didn't have a brain that was capable of overthinking literally everything, The Fuck It Diet would be so easy. If you were a groundhog, for example, and just worked on instinct, you would have followed your hunger, gotten out of a low metabolic state, and that would be that. Animals don't have body-image issues. And they don't let themselves get into guilt-repent cycles with their eating.

You would have easily let your body eat a lot to compensate for your famine/diet, and you'd have let your body gain weight. You'd have let yourself spend time resting and repairing, and boom, before you knew it, your famine/diet and recovery phase would be a distant memory. Your appetite would normalize, and you'd be able to focus on a million other things.

But that's not what we do. We overthink. We panic. We think that it's all going horribly wrong. We let our *beliefs* trigger emotions and panic, which we are then too afraid to feel. We doubt ourselves. We doubt the process. We doubt our bodies. We worry that we will become unrecognizable and miserable and unlovable. And we spiral into more panic.

START COLLECTING BELIEFS

When you do your brain dump, start circling your beliefs and write them in a separate list. Start taking note of beliefs that might be running the show and keeping you stuck. In fact, really complaining and writing about what's upsetting you in your brain dump can help you figure out what beliefs might be causing it even more.

You can also ask yourself, "What beliefs are making me feel this way?" Then write it out.

WE ALL HAVE OUR REASONS

While I was in high school and began gaining and losing weight, I was vacillating between bra size E and H. And at the exact same time, Jessica Simpson was in the tabloids yo-yoing too. Jessica would lose weight on a diet, then months later would gain all this weight back in her boobs, arms, and face—and the tabloids would rip her to pieces. And all I could see was: *That is my body. That is how I gain weight and lose weight too. I'm a mess too. I should feel ashamed too.* The way the tabloids ripped her up, I ripped up myself. And she was *beautiful.* God, how ugly must I be?

In the span of a few months in eighth grade, I went from being able to go about my life feeling like a happy-go-lucky child to having trucks honk at me as I ran, and creepy men catcall me on the street as they drove by. I couldn't believe it was happening. Who were these people? Was this normal? Did they not think I was a person? I was still in middle school, and this became a constant reality for me. I linked it to my boobs, which was linked to my weight. All of a sudden, my body was the thing that defined me, and I had no control over these open, aggressive reactions. In my teenage brain, the harassment felt like punishment for not being skinny anymore, and for a very long time I thought that losing weight would make it stop.

But street harassment wasn't the only reason my relationship to weight became so dysfunctional, because right around this same time, when I was fourteen years old, doctors told me not to gain weight and to watch my carbs and fat intake because of PCOS. For me, this was just more proof that weight loss and weight fixation were an important and legitimate cause. Nobody could convince me that I was becoming disordered with food, because that relationship was basically doctor-ordered.

To top it all off, I wanted to become an actor. I wanted to go to college for musical theater and make acting my career. In a weird way, my burden was that I was naturally really talented, and so *if I could make sure that I looked right*, I could get into a top program. If I looked tiny and dainty enough to match my pretty little voice and "type," I could make my dreams come true. But if I *couldn't* master my weight, I felt like I was committing self-sabotage. I'd never get into an elite program, and never be hired. *And my boobs don't fit into ANY DRESSES! This is all my fault. I need to go back on the Atkins diet and get some fucking control once and for all.*

I believed that if I wanted to follow my dreams, stop being harassed on the sidewalk, and not have my health spiral out of control into diabetes and infertility thanks to PCOS, I *had* to lose weight for good. I had to work harder on my willpower. I *had* to stop eating carbs and stick to the diets that were becoming harder and harder to stick to. And if I couldn't, I was letting my food addiction undermine my *destiny*.

I was convinced that my health and career were actively on the line, every day, every bite I took, every pound I gained—and that my weight was the cause of all my problems. IT FELT VERY DIRE. Every bite felt like a test: Could I be a successful human or not? Could I remain worthy, successful, and admirable, or would I spiral into disease and ugliness and failure?

Meanwhile, as my body fought back against my obsessed attempts at willpower and restriction, food began to take over every thought I had, and my shame kept deepening over my inability to successfully stick to a diet. *WHAT is wrong with me? Am*

I truly this broken and out of control? It was absolutely miserable. And nobody was able to tell me that what I was doing and believing wasn't healthy, because we are living in a culture where most people are doing the same thing to varying degrees, and being cheered on for it.

That was my formative experience. You have your own. We all have our own series of events that made us believe we needed to control our weight. We came to believe that our lives would be better if we could become smaller or fitter. It may have been something someone said once, or something we heard repeated over and over. It could have been something seemingly common and innocuous, or more textbook traumatic.

Examining the experiences that led to your weight and food fixation isn't meant to keep you dwelling in the past, or stewing over coulda woulda shoulda; instead it will help you begin to examine the core *beliefs* that those experiences formed. Those are the beliefs that are probably running the show from the background, alive and well in your subconscious, but aren't actually serving you at all.

For instance, thanks to my experiences with catcallers, I internalized the idea that my weight made me unsafe, and that my boobs gave me no control over how people treated me. And that being curvy made me fair game for disgusting or aggressive comments. *And* that being skinny was the one thing that would make me respected or keep me safe. None of those beliefs are helpful. Also, none of those beliefs are inherently true. But I was making them true by holding on to them.

WRITE YOUR STORY

Take a good twenty minutes to write out your food and body story up until this point. Write out what life was like *before* you started dieting, how and why you started dieting, what led you to believe you needed to diet and lose weight, and what it felt like while in the trenches. Really go there. Remember things you've willed yourself to forget. This will probably be a miserable story and may be uncomfortable during the process. This is the first step in healing. Bonus points for being in your body, breathing, and feeling while you write and remember.

Next, go through and underline the specific experiences that clearly turned into limiting beliefs that are still affecting you today. For instance, remembering "I kept getting praise whenever I lost weight" might have made you believe that "losing weight makes people proud of me."

Add to your separate list of these limiting beliefs and save this list to reference. We are going to keep collecting our negative beliefs to do further work on.

Remember, the more you can breathe and feel during this whole thing, the more you begin the untangling process.

WHAT WE THINK BEING SKINNY WILL GIVE US

Here was the plan: once I really nailed the low-carb diet thing and never ate bread and became permanently skinny, I'd dye my hair blond. *Then* I'd be thin *and blond*—and maybe even get a nose job. But maybe shrinking my big cheeks would help with the nose somehow—we'd have to see.

When I'm thin I'll be beautiful and confident and hired for all the theater jobs I audition for. I'll go to parties all the time. I'll love socializing and laughing. And I'll be funny and cool and I'll dress well and everything will just be so so so much easier.

I also had plans to be in a big cultural-political drama by somehow being in a play with one of Prince Harry's friend's wives in London, and then we all would go out to dinner, but I wouldn't really know who he was, because I didn't really pay attention to the news, and he would fall in love with my grounded, charming, beautiful, thin, blond, oblivious, non-power-hungry self. It's also important to note that when I started writing this book, Prince Harry was single, and now, as I am editing it, he just got married to a charming, beautiful, thin, *not*-blond actress. Oh, how our dreams die. I had lots of other fantasies that happened in London too, a lot of them about how I would meet George Weasley, preferably in a Muggle coffee shop.

The common denominator in all of the fantasies was that I would be thin. That was the most important one. The other common thread, which I didn't intend, was that I was going to be in love with a red-headed British man of high esteem. (George has a *very successful joke and candy shop for wizards and witches.* Though clearly I wasn't planning on eating any of the candy from his magical candy shop because I was a carbless goddess.)

Through the years, depending on the diet I was on, the fantasies shifted and morphed. During my *French Women Don't Get Fat* phase, I was going to be thin with an amazing wardrobe of slinky, high-quality staples and silk shirts where I wouldn't need to wear bras 'cause my boobs shrank so much. I would only have what I needed. And my apartment would look like Amélie's.

There was a time in high school, around when the movie *Chicago* came out, when I was truly depressed that I didn't look like Catherine Zeta-Jones. I was miserable about it for *months*. I spent countless hours looking at pictures of her on this new website called Google. I'd stare at her perfect face and freak out over how very *not like her* I looked. And *she* was on the Atkins diet! Just *more* proof that eliminating bread from the planet was *my only hope*.

Lots of us fantasize about a life where we lose weight, get what we want, and everything falls into place—maybe sans wizarding-world fantasies, and yes, I will admit that my imagination is *very* elaborate—but still. Losing weight is a fantasy that our culture actively perpetuates, and advertising companies will use your fantasy to get you to buy their pointless shit.

Our deep desire to be thin or fit is implanted in us by society, and then exploited until we spend all our money chasing the dream of what thin will give us: Happiness. Respect. Love. Confidence. Beauty. Nice watches. A very white kitchen filled with tiny yogurts. Relaxation. Peace.

Most of us buy into this fantasy without ever realizing that we have—it just feels like a truth. But that subconscious brainwashing is what we can change right now. All you need to do is start becoming aware of your thin fantasy. And then start allowing yourself to have the things that you think come along with your thin fantasy *now*. And fuck anyone who says you can't.

Do you think that anyone was going to tell Lena Dunham, the creator and star of the HBO show *Girls*, that she was the perfect person to star in a TV show? Fuck no. She gave that right to herself. She made her own movie and then her own TV show. She changed the game. I don't care if you love or hate her, or the show—she did what you can do too. She didn't wait to be skinny to do the things she wanted.

The belief that being skinny will make us happy sets us up for failure. External things are famously unable to make us truly happy. And still, so many people's entire lives revolve around the belief that being skinny (or fit or lean or whatever) will make them happy. Or being wealthier. Or being adored. Those beliefs will affect the way you treat yourself, and once you get what you think you want, you'll be even more miserable when you realize that it didn't create the lasting happiness you expected.

I am here to tell you, and then tell you again, that everything you think being thin will give you is something you need to be willing to seek now, regardless of your current weight. You were not made to sit around waiting until someone deems you good enough for the life you want. You were made to go create it.

We all think that once we become skinny, or beautiful, or rich we will finally be happy. But the truth is, the way we seek things out is the way we will experience them. If we seek out a relationship in insecurity and neediness, we will most likely spend that relationship in the same state, still insecure and needy and seeking constant validation. Same thing with weight loss. Losing weight doesn't cure the emotional state you're in. Losing weight doesn't change the way you feel about yourself or the beliefs you have about yourself. I know that sounds counterintuitive, but this happens all the time. There are so many people who lose lots of weight, and even if they are now getting endless praise, *they still treat themselves the same way they did before.* They still don't like themselves.

I'm not anti-goals. I've got goals. But I *am* anti-expecting-external-goals-to-actually-make-you-happy. That raise will not solve all of your problems at work. Falling in love does not erase self-doubt or feelings of loneliness. Fame and recognition does not automatically make you peaceful and content. We have to look at what we are really searching for *underneath* the goal. If what you're really seeking from weight loss is more kindness to yourself and a cute new shirt . . . you need to be willing to give those things to yourself *now.* That's the underlying stuff you're actually seeking. Because there is no guarantee that getting to your goal will make you kinder or more accepting of yourself, even if you assume it will. Most often, it won't. The way you seek out a goal is the state you will still be in once you get there.

It doesn't mean none of your goals are worth it. It just means that your goals won't automatically give you the feelings you are seeking. Expecting the goal to make you happy will only end up disappointing you. But the good news is, if you can identify the feelings you are really hoping your goal will give you, you can experience them even sooner than the goals themselves. Because happiness and "real life" aren't going to happen *then*, or "once you finally . . ." Life is fully and always happening now.

And I promise you that you will look back on your life *now*, one day, and wonder why you kept waiting for later.

GOAL EXPLORING

Write out five of your top goals. Don't let it stress you out, just jot down some goals you've had, even if they are shifting.

Once you have your list, spend some time figuring out why these are your goals. What do you want to experience when you attain them? How do you hope to feel when you get there? What kind of things will you do and think? Great job. Work toward giving yourself permission to have those things and experience those feelings now. Before you even reach your "goals."

LOSING YOUR IDENTITY

I have heard readers ask me something like this again and again: "I don't think this will work for me. Working out and maintaining a low weight are some of my core values. How am I supposed to be happy if I sacrifice my core values?"

Maintaining a low weight is one of your core values? Like treating others the way you'd want to be treated, or being honest? Maintaining a low weight is not a core value. It's a fear-based ingrained societal standard, created to make money off your insecurities. Weight control relies on fear and fixation. The thing we like about it most is the high of fitting in, getting praise, feeling safe, and the temporary relief that comes when we reach a goal weight. *Whew, now everyone will leave me alone and approve of me. Now I'llllll leave me alone.*

That's until it isn't good enough anymore, or until we gain it back and feel horrible about ourselves, and the cycle of shame continues.

Health, movement, eating what feels good, and dressing yourself in clothes you like *aren't even core values.* They are, however, awesome ways to take care of yourself. Feeling healthy and strong and embodied is a perfectly legitimate *desire*, but living in a constant food and weight obsession is not.

"Staying healthy and thin" as a core value also relies on health and weight being fully within your control, and the assumption that controlling your food and weight will actually lead to better health—all things that you can't account for, and have even been proven untrue.[65] Goals and core values that are more self-loving and self-forgiving will probably end up being better for your overall health anyway.

What is completely understandable, however, is the adjustment period of losing your old identity and having to figure out who you are without it. Who are you without the goal of being skinny?

Who are you without being able to take on the label of "the healthy one"? What do you actually *do* with yourself during the hours you used to spend meal-prepping low-fat, low-carb food for the week?

A core value that'll serve you better as you try and piece together a new, more forgiving identity could be "prioritizing your needs" or "taking care of yourself." *And* if you have a weight obsession or any disordered eating, prioritizing your needs is gonna look a lot like The Fuck It Diet.

You have every right to remain someone who judges your daily worth based on your weight, but it's not gonna be fun for very long.

MOURNING THE OLD FANTASY

Take some time to remember that fantasy you had of the person you wanted to become. Without judgment, and understanding that you truly just wanted to be happy, allow yourself to honor and grieve the fantasy of the body and life that ended up not really serving you.

WHY WE PANIC

If you believed in your core that you were taking care of your body the best way you knew how, and that you were listening to it to the best of your ability, and that feeding yourself was the best way forward, and that you were beautiful but also completely undefined by your looks, and that your power lay in your heart and your creativity and ingenuity . . . then when someone made a comment about how big the sandwich you were eating was, your response would easily be: *Yeah! It is a big sandwich!*

Or when you didn't fit into the pants you used to wear, your reaction would be: *Yep! Humans fluctuate and I am healing and listening to my hunger the best I can. Let me get new pants.*

Or when your close friend or family member expressed concern over your higher weight and "health," your response would be: *Oh thank you for caring! I understand why you're thinking that way, but I am SO much happier and saner than I was. I'll continue listening to my body, and I'll update you along the way. I feel really good.*

But instead, we panic. And the reason you're going to respond with panic and tense shoulders instead of with a shrug is because your beliefs are falling in line with society's beliefs about food, weight, worth, and health instead of more empowered (and *truer*) beliefs. Very physically, when our core beliefs feel threatened, the *same part* of our brain that responds to physical threat gets activated, which in turn activates the fight-or-flight response.

So when you secretly believe deep down that you really *are* being irresponsible, and that you really *are* worthless and ugly . . . YEAH! You'll panic! If you secretly believe that when your clothes don't fit it's a sign of your deep moral failing, and that everyone is disappointed in you (and that they *should be* because this is all your fault)—*YOU WILL PANIC.* And you'll panic when *any* little thing happens to trigger your fears and beliefs. Of COURSE.

But here is an important reframe: All of your little panic meltdowns are also gifts, because they are pointing to limiting beliefs that *you* need to recognize, become aware of, and *let go of.* If you're panicking, there's a limiting belief that you can find and become aware of.

THE AVOIDABLE KIND OF PAIN

Life is painful. I'm not able to offer you a pain-free life. That's not the deal here on planet earth. But it's important to understand that *some* pain is being made *worse* because of our beliefs.

Unavoidable pain is mostly grief-based. It's the pain you feel when your heart is broken, or when your family member or loved one dies, or when you lose something else you cared about (like the fantasy of thinness curing all our woes). We grieve. It's a part of the letting go. It's a part of life. And it is part of being human. When we lose things or people we love, we *need* to grieve.

Similarly, when we are treated badly or experience emotional trauma, we hurt and we grieve. If we are willing to feel it and able to process it, that grief allows us to learn about ourselves, and to honor the change and loss.

You cannot go through your life without losing and grieving, and if you try to avoid it, that emotion will get stuck and be constantly waiting for you to feel it. My advice on unavoidable pain is the same as always: feel it and honor it. That's how you process anything. Come back into the body to feel it, use the breathe-and-feel tool, and it will pass in time. It will come in waves. It will teach you how to be human. It's not necessarily fun, but it's so incredibly important. And paradoxically, feeling grief and pain will eventually allow you the space to process, and eventually be happy again.

The other kind of pain is belief-based pain, and it is *way* more avoidable . . . once you learn about it. This is stress that you will feel because of your beliefs, and in our case, beliefs about weight and how you *should* be and *should* look: *I'm unacceptable. I shouldn't look like this. I'm a total failure. My body is disgusting. Everyone is judging me. Everyone is right to judge me*, and on and on.

Those beliefs are causing most of the emotions and misery you may be feeling. Much of it could be avoidable if you could instead say to yourself, *Um, I'm awesome and doing the best I can—and you, who are trying to shame me, are clearly a confused asshole.*

If you believe your stomach is unacceptable, you will be miserable over it. And that's belief-based pain. But if you changed your core beliefs about your stomach, you'd get rid of that specific trigger of fear and insecurity, and experience *less* pain over your body.

But say someone makes a rude comment about your stomach . . . that could cause both unavoidable *and* avoidable pain. You might still be pissed or hurt that someone is pushing their shitty beliefs on you, or feel real grief that you have to live in a world where people *make fucking comments about people's fucking weight.* That's grief-based. That's unavoidable. But depending on whether you *believe* them or not, that can change how painful the whole situation is. If the rude comment doesn't line up with your beliefs about your stomach and your worth, it won't have power over you in the same way. You'll be able to see through the bullshit, and *that* is our goal.

And the more unexamined beliefs you have, the more potential there is for a snowball effect. For instance, when someone makes a rude comment, you could also spiral into *This must be what everyone thinks. It matters what they say. This proves I am disgusting. I can't trust anyone. I should listen to them. No one respects me.* And on and on. These lurking beliefs are going to make the pain even bigger than it needs to be.

But if you didn't have these beliefs, you wouldn't be so upset. The pain would stop at *What the hell is wrong with them?* and it wouldn't spiral into misery and panic and pull that big tangled knot even tighter. Gaining some awareness around which beliefs are snowballing into these emotions will help you stay in the helpful, unavoidable pain and minimize the avoidable, belief-based pain.

STOP SHOULDING ALL OVER THE PLACE

By now you've probably noticed how much misery we create and perpetuate by thinking that our life is supposed to be some other way.

That if only I was better, thinner, older, younger, richer, in love, prettier, funny, outgoing, smarter, better employed, I'd be happy. Then everything would be better.

And I call that *shoulding.* Our brain will verbalize our mental blocks as either limiting beliefs—"Fat is bad"—or shoulds: "I *should* be thinner."

And *all* shoulding is some version of "This shouldn't be happening" or "Something else should be happening."

Here are some examples of shoulding:

> *I should be further along by now.*
> *I shouldn't be struggling with food.*
> *I shouldn't be this size.*
> *I shouldn't eat like this.*
> *I shouldn't be tired.*
> *I shouldn't crave sugar.*
> *I shouldn't be single.*
> *I shouldn't be unhappy.*
> *I shouldn't be struggling with my career.*
> *I should be craving healthier foods.*
> *I should have lost weight by now.*
> *I should be eating less.*

Very sincerely, shoulding is ruining your life. *All* shoulding is *the worst.* And yes, shoulding *does* sound like shitting when you say it out loud. You are right.

We think that shoulding is a responsible way to live, and that we won't "improve" unless we shame ourselves into doing better. But what it really does is put you in a spiral of shame and guilt that is impossible to get out of. Similar to the binge/repent cycle, but this is a mental and emotional cycle of doom.

Shoulding is going to really pop up as you go on The Fuck It Diet. People have a very specific way they think this thing is supposed to look. They hope that they can allow food, maybe eat a lot for a few days, and then become soooooo chill and normal around food immediately. Then they'll be craving asparagus and mahimahi and maybe a peach for dessert by week two. By week three they are miraculously really thin, and when people ask how they did it, they say: *It's amazing, I stopped dieting three weeks ago, I eat whatever I want, but isn't it amazing? I only want fish and veggies anyway!*

I am here to gently remind you that that is almost certainly not how this thing is gonna look. Mahimahi and peaches are great, but there is a reason why I am continually reminding you how LONG it would take to heal after a year- or decade-long famine. There is a reason why I tell you that you cannot *think* your way through this. There is a reason why I am spending lots of time explaining how important it is to accept your weight at a higher size.

Shoulding is going to be the reason you are unnecessarily stressed. *But wasn't this supposed to heal me? This should have made me crave fish and veggies by now. I should be further along. I should be doing this all differently.*

You'll be miserable if you believe that it is supposed to be happening another way. Basically, stop shoulding. It's causing way more stress than you have realized. When you're miserable, look for the limiting beliefs and shoulds that are causing the anxiety. You can treat shoulds the way I teach you to treat all limiting beliefs.

ALL OF YOUR SHOULDS

Now make a list of all of your shoulds, not just about food and body, but about everything you should over.

For instance, I should make time for ________. I should have done ________ by now. I should have figured out ________ by now. I should be doing ________ differently. And on and on.

Let this list be really long. Write until you cannot think of any more. There may be overlap with the limiting beliefs you have already found—that's okay.

These shoulds are all things that are pulling at you. Just starting to be aware of them is extremely helpful.

YOU CAN RELEASE YOUR LIMITING BELIEFS

Like we went over in the emotional section, unprocessed emotions and energy will get stuck in the body. When you use the breathe-and-feel tool to get into your body and feel sensations and emotions, you may notice that images, memories, and realizations come to mind. That's because certain beliefs are *tied* to different emotions and stuck energy. Which means you can also access this emotion by *starting* with the limiting belief and using it as a way to access and process the emotion and energy tied up in that stressful belief. That's what the next tool is going to teach you to do.

For instance, a very common belief is "I should be ashamed when I gain weight." That belief will be attached to a well of energy, emotions, pain, memory, and shame that we do not want to have to feel.

That energy and emotion has been walled off, so to speak, and not dealt with. We've created energy walls within our body so we don't have to feel or deal with the feelings the belief brings up. Anything can trigger and bump up against those walls, but our habit is to feel that initial discomfort and then try and wall it off even more, which just causes more stagnation.

Tool #5 is kind of like a brain dump + breathe and feel. You will be using a stressful belief as a prompt to help you activate and access the energy associated with it in your body. You're going to lean into the energy by writing about it, and while you do that, you're going to feel what is coming up by breathing into it. The goal, as always now, is to feel what you usually run away from feeling, using the breath to actually activate it *more.*

These are some big limiting beliefs that often accompany a disordered relationship to food and are worth addressing as you take yourself through The Fuck It Diet:

I am a food addict.
Being full is unhealthy.
I just need more willpower.

It's all my fault.

I don't deserve to relax.

I need to be responsible.

It is not safe to feel.

If I start to feel, the pain will never end.

If I am not thin, I am failing.

If I am not thin/beautiful, I will ______.

I don't deserve to take care of myself if I'm fat.

I cannot accept my current weight.

Being fat is ugly.

Only thin people can ____________.

I need people to approve of me.

Gaining weight is unhealthy.

Gaining weight is ugly.

Losing weight could be easy if I had willpower.

Losing weight is responsible.

Nobody will take me seriously if I gain weight.

Most food is bad for me.

Gaining weight is a sign of failure.

My weight is my fault.

I can't trust my body.

I should limit carbs.

I can't eat a lot.

Being skinny will make me happy.

» TOOL #5: THE BELIEF RELEASE

» Choose a limiting belief to work with. It can be one you found from your writing, or one that is listed anywhere this book.

» Find a quiet spot where you won't be interrupted and get a notebook or pages you plan on burning or shredding or recycling or whatever your little dramatic heart desires.

» Write at the top of the page: I am releasing the belief ________________.

» Start writing anything that comes to mind about this belief. Memories. Emotions. Side thoughts. Where the belief came from. Why it feels hard or scary or impossible to let this belief go.

» As you write, notice where you feel stress or discomfort rising up in your body: legs, low abdomen, mid-abdomen, heart, throat, anywhere you feel stress or sensation. Breathe and feel into it. That's really all we are going to do: continue to activate stuck stress, and breathe and feel to process it. You may feel a lot, or you may feel just a little. It may be clear emotions, or it may be more like tension and sensation. It doesn't matter what it is—give yourself permission to sit with that feeling. Feel what you usually run away from.

» If you run out of things to write about, but still feel like there is more to feel, focus on writing about where the belief came from, and why it's hard to let go of, and keep breathing into what it brings up in the body.

- You get to decide when you've had enough. Even a minute or two is time well spent. You could also wait until you feel any sort of shift, and the stress doesn't feel as easy to access anymore, because that means you probably did a lot of good work. But small, bite-sized chunks work very well and are a good way to start.

This isn't about what you write. This is about what the writing allows you to *feel*. Sure, you may come up with major breakthroughs and memories you had previously forgotten about, but the way to release them is the feeling part. Your only task is to breathe and feel what is there. Have I said that enough times yet???

What you write or remember does not have to feel epic in order to be profound. You can write mundane things the whole time, breathe and feel, and still release lots of stuff that's been getting in your way.

And remember: be kind to yourself. Processing energy can be taxing. Rest, eat, try an Epsom salts foot soak, or do anything else that feels grounding or replenishing as you work through these feelings and beliefs.

NOTHING IS A CURE-ALL

Tool #5 is a form of "energy work," and there are so many methods of energy work. Some other energy work is meridian-based, or more chakra-based, or muscle-based, or more pressure-point-based; some of it is directly on the body, and some is not. Some is focused on healing ancestral trauma passed down from past generations. Some is more spiritual, some is more body based. If you are interested, try as many things as you like, or keep it as simple as I've made it.

I have trained in energy work methods alongside psychologists who said they were "in the closet" with their energy work, but confessed, "I can make SO much more progress, so much more quickly, with my clients when I incorporate energy work with the ones who are open to it."

Energy work is a tool that works alongside all of the other information about eating, dieting, weight, and health that I've shared up until now. Use it however you see fit. The big takeaway is: feel.

But it is important for me to remind you, as someone who is expressly against diet and spiritual dogma: **Nothing is a cure-all.** Nothing. Energy work is about being willing to dig deeper. Coming back into the body and feeling emotions can help people process and move past things that have felt impossible to move past. Energy work can guide you through feeling those feelings. It helps because we finally become willing to feel things we have avoided feeling.

IT WAS NEVER YOUR FAULT

I carried the belief that this was all my fault around for a long, long time. *This is all my fault. My weight is my fault. My health is my fault. I'm tired and that's probably my fault too. I'm bad at auditioning, that's totally my fault. I'm ugly and it's all my fault. And when I die alone, that'll definitely be my fault too.*

At a certain point, I realized that for everything I was stressed about, I was actually less stressed about the actual thing, and *more* stressed that it was probably all my fault. I was always worried that I was supposed to be *doing something* about it—and that I had no one to blame but myself.

My student Charlotte was walking around for years with horrible pain in her foot. Every step she took was painful, and she was sure it was because she was too fat. She was blaming her pain on her weight. She believed if she was skinnier, and stronger, her foot pain would go away. She didn't even need to ask a doctor about it, because she was so sure they would just tell her to lose weight.

So for years she would diet, and she would lift weights, and try to lose weight—but it never cured her foot. Finally after *ten years* of just accepting that her foot pain was on her, she went to a doctor who simply said, "I can't believe you've been walking around on that. This is a damaged nerve in your foot called a Morton's neuroma, and we can surgically remove it. I can do it tomorrow, in the office, and it takes about twenty minutes. You can walk out of here in a surgical shoe."

Morton's neuroma has nothing to do with weight, and people of all sizes can have it. In fact, the high level of activity she had maintained trying to "cure" it had actually made it worse. Charlotte had spent so long assuming that her pain was her own fault, and that she wasn't trying hard enough to take care of herself, and that she "should" be able to control the pain, that she never even questioned it. She now walks pain-free on her foot, no dieting or weight loss necessary.

Some things are our fault. If we are perpetually rude, yes, it's our fault when we begin to treat people poorly. Taking advantage of people is our fault. Doing things to hurt others is our fault. But weight, health, and looks? Bad luck? Financial setbacks? Health? Those things are not fully within our control, and you *cannot* hate yourself into improvement.

We were all taught that if we tried hard enough, bought the right products, and devoted our lives to "healthiness" and constant productivity, we too could become more beautiful and more acceptable, ward off our impending mortality, and be finally happy. And so that is what we tried to do, as dutiful members of society.

In some ways we do have control over our health. Sunshine, water, nourishing food, sleep, circulation, relaxation, embodied and joyful movement . . . those are things that are great for your health.

Some people can improve their health a hundredfold by adding in some yogurt; some people are born with mercury poisoning and simply have a harder go of it. If your health is a stressor and you have blamed it on your eating or your weight, or now are even blaming it on your dieting—it's not your fault. You were always doing the best you could with the information you had, and some things are really, really hard to figure out. Some things will take your whole life to figure out. Some things cannot be figured out at all. Sometimes, surrender is the real lesson underneath it all.

I truly believe that most people are doing the best they can, even when that best doesn't seem very good. Even heavy drinking, smoking, drugs, constant avoidance and numbing . . . People are hurting, and are *still* doing the best they can with the coping mechanisms they have. And once they are ready or able to move on, they will.

Let yourself off the hook. Unless you are a rapist or a racist. In that case, I'm not talking about you. Go get help. You are an asshole.

DIET CULTURE TRAUMA

Fat shaming is trauma.

—JES BAKER

Being shamed and berated for the size of your body is traumatic. Just like we talked about in the emotional section, we can get truly traumatized on a very physical level from seemingly innocuous situations. It's a survival response. And think about it: up until relatively recently, we relied on our tribe or community to survive. Being ostracized used to be a very real threat to our survival. We have a deep desire to be accepted, not only emotionally but also on a very primal level.

So we are treated cruelly, and then, after the initial trauma, we're expected to diet and try and lose weight, which forces our bodies into *another* survival state. That is *even more traumatic.* And then to have those diets fail and blow up in our faces over and over, for years on end, no matter how hard we try? It's overwhelming. It's disheartening. But even more importantly, it's *traumatic.* There is a buildup of emotions and energy associated with these experiences, which is why gaining weight causes such major panic: we live in a society that tells us that gaining weight is the last thing we should do, and the worst thing that could happen to us, where people are openly ridiculed for gaining weight or being fat in the media. It's no wonder we are all so emotional and anxious and hard on ourselves.

In the YouTube comments on a video I once posted about The Fuck It Diet, a man commented that body positivity was part of the "absurd liberal agenda to let everyone feel good about themselves," which is apparently supposed to be a very bad thing. Enough people really think this way, though. We're all supposed to be walking around feeling horrible about ourselves, repenting for our existence by feeling as much shame as we can?? That doesn't work.[66]

Being shamed, or witnessing *other* people being shamed, or being told that they *deserve* to be shamed based on their weight because it is all their fault leads us to develop coping mechanisms to try and avoid this pain in the future.

My student Jenna reached out to me and told me,

> I was truly having a breakdown from being surrounded by diet culture and fat phobia. My therapist called it PTSD. WTF?!? At the time I thought that was too extreme of a diagnosis, but now it makes SO much sense. I was traumatized. I literally could not talk about food, or weight, or dieting, or listen to people talk about it, without bursting into tears—which is difficult when diet talk is SO pervasive. It is EVERYWHERE.
>
> But now something has shifted in a huge way after finding The Fuck It Diet and the energy work. I've only

> tried it on maybe four limiting beliefs—but I feel a major difference. Mostly that I can talk about being anti-diet without being a blubbering fool, and I can eat without guilt and I'm no longer bingeing—what?!? I am accepting my body more every day, which is new to this fifty-three-year old who has hated my body since I was a teen. It has truly been a transformation.

There is a magic in the combination of trusting your body and appetite *and* beginning to get back into that body and feel and process the things you have been avoiding. Even just having an understanding of *why* you can't stop crying, or *why* your heart begins beating really fast when you think of weight or diets, can help you navigate this whole thing with more kindness toward yourself.

Many of you might need significantly more individualized counseling, therapy, and guidance to work through your trauma. But these tools and concepts are a good solid start. Feeling will never be a bad habit, no matter where you are in your journey. This energy work I share with you is a starting point.

THE LUCK OF THE THIN

How many of these apply to you?

- People assume you're unhealthy because of your size.
- People comment or judge the food you eat in the name of *trying to be helpful*.

- When you go to the doctor, they tell you to "lose weight first" instead of just treating the issue.
- Your health insurance rates are higher and airlines charge you extra to fly because of your size.
- You're less likely to get a raise or promotion at work than someone who is thinner.[67]
- You're the brunt of constant jokes about fat people stuffing their faces.
- The media describes your body shape as part of an "epidemic."
- You can't find clothes in your size in major stores.

If you are fat, you know all too well that you are *not* benefiting from the luck of the thin. You do not have the privilege of going through life with people ignoring your size. You feel the constant judgment, the stares, the eye rolls, the rudeness, the coldness, the blame. You may dread going to the doctor, flying on planes, or just having to go out into the world to face the large number of people who think *they know something about you*—how you live, how you eat—or push their opinions on you disguised as concern for your health.

Thin privilege is a spectrum, too. It's interesting for people who have never felt particularly *thin* to learn they are still benefiting from some thin privilege. Even if you aren't actually "thin" at all, but fall into the medium weight range, you can still benefit from thin privilege. For instance, even at my highest weight I still had thin privilege: I could find clothes in my size in any store. Doctors don't blame my health problems on my weight. And I was generally able to go through my life avoiding the judgments and assumptions that come with having a fatter body. And even now, I am writing a book about eating and gaining weight, and I get to benefit from the fat-positive movement while *still* being able to experience and benefit from thin privilege. That is why it's so important to hear from people who *don't* have thin privilege too (more about that in the next chapter).

When you're thin, undereating and overexercising are seen as disordered and dangerous. But fat people who are doing the exact same thing are seen as just doing what is "responsible and necessary."

Many fat people actually have anorexia—putting their bodies through extreme restriction, as they are encouraged to do—but they still don't *look* like it, because their weight set range is high. A weight requirement for diagnosing anorexia is becoming old-school thinking; *behaviors* define eating disorders, not body size. Both thin and fat people will experience the same hormonal starvation mode; the only difference is the external weight set point.

One of my students told me she got to a point where she was only eating vegetables and fat-free yogurt for breakfast, lunch, and dinner, and had lost significant amounts of weight, but her BMI was still in the "overweight" category. She didn't *look* like she had an eating disorder. She was also experiencing all the other signs of starvation mode: low body temperature, dry skin, trouble sleeping. But her doctors just praised her weight loss, didn't ask what she was doing but told her to keep it up, and even told her that her dizzy spells and absent menstruation were because she still wasn't *thin* enough, not because she wasn't *eating* enough.

We live in a society that praises fat people for being obsessed with losing weight, even when they develop eating disorders in the process. The extreme measures that fat people are expected to put themselves through—*supposedly for their health*—measures that are dangerous, unnatural, and have many known health complications, is completely hypocritical.

For instance, weight-loss surgery induces a starvation response, which often causes quick and severe weight loss, *plus* malnutrition—and ongoing, an impaired metabolism, which often leads to inevitable weight regain and poorer health, despite the surgery. But hey! Our society knows what it values: weight loss. *Anything* to lose weight.

People are sometimes resistant to the idea of privilege, because they fear that admitting privilege will invalidate the hardships they *do* face. But that's not how it works. You can have privilege and still have problems. Your life can still be hard even if you were born with certain privileges, like having a naturally thinner body, or being born white, straight, rich, or whatever. All of those privileges are *luck*, luck of fitting in with ease, and they come with benefits that other people don't experience—and that we often take for granted. Becoming *aware* of the things we have taken for granted will help us create a kinder, more empathetic and aware society.

Health has so much less to do with our own habits and so much more to do with our social and financial positions.[68] How oppressed are you? How stuck and unseen do you feel? How difficult is it for you to make ends meet and make things work?

How hard is it for you to take time for yourself? To breathe a sigh of relief? How marginalized are you by society? How powerless do you feel? How hated do you feel by other groups? How unsafe do you feel? How much are you taught to hate yourself and blame yourself?

There was an experiment in the 1990s that gave diabetes patients housing vouchers, and their symptoms improved just from better housing.[69] Not from health care. Not from drugs. Not from exercise. But from a better life experience and less stress. A different study found that institutionalized children with access to similar diets grew at different rates, depending on whether they were tended by warm or stern caregivers.[70] How we are treated *matters.*

So when fat people are treated like shit, and develop stress-related health problems, they are told that their health problems are their own fault because of their weight. They are shamed and overtly told to diet, but the shame and stress *is doing the most damage in the first place.* This sets up a terrible cycle of dieting, stress, and health problems. There is no winning in this paradigm. And, my friend, this is something green juice cannot cure.

FAT PHOBIA

So if our feelings of power and autonomy make such a difference when it comes to health, why is this not more common knowledge? Well, if it was, we wouldn't be able to blame individuals for their health or weight problems. We would be forced to implement social reform, and finally admit how important quality of life, kindness, and inclusion are in the bigger health equation. Because how can people rise out of their current social and financial situation if that situation is the very thing keeping them sick?

I didn't always think this way. I just didn't know better. I really thought that being skinny was the *only way* to be happy, and could become fully in my control. I thought becoming skinny was the only way to be beautiful, acceptable, and successful. Worshiping skinny may as well have been my religion. I internalized our culture's fat phobia. And not only did I turn that fat phobia in on myself, but I was judgmental of other people too. *Well, at least I am doing better than them.* We tend to think that way when we are scared and insecure, and it's bad, bad, bad, and I am retroactively sorry.

Part of my Fuck It Diet epiphany was about weight: *I need to accept my weight wherever it falls, or I'll never be happy.* I knew it deep in my bones. I didn't know how to do it, but I knew it had to happen. Even still, I was resistant and petrified of weight for a while, no matter what I consciously *wanted* to feel.

I didn't realize how *deep* fat phobia runs in our culture. It is so rampant that we often don't even realize we are a part of it, but this widespread fear of weight and fat affects everyone. We *all have it.* We are afraid of being fat whether we are fat or not.

One of the most healing things I did was follow fat activists on the Internet, read their books, and hear their stories. I started following the accounts of people who were proudly fat and sharing their stories, experiences, and pictures. Some of these activists are athletes, or models, or writers, and the majority of them have already spent decades of their lives hating their bodies and trying to lose weight, to no avail.

Seeing people of different sizes choosing to be happy and beautiful and confident, even though they were always told they couldn't be because of their weight, was the best way to unlearn what I believed about weight and worth and happiness. They are living examples that being fat doesn't have to mean what we've been taught it means. *You* don't have to feel the way you've been taught to feel. It's a reminder that it really doesn't matter what some people think of you. It matters what *you* think.

This isn't to downplay how fat-phobic our culture is. It is hard to live in such a fat-phobic world, *even when* you decide to love yourself. Living in a world that is openly afraid of fat will bring the unavoidable kind of pain. And it is significantly harder for fat people than thin people, because thin people get to benefit from fat activism and never have to deal with the active judgment.

I recommend you follow the many brilliant fat activists who write from their own perspective on how to deal, survive, and thrive in a world that is habitually cruel and oppressive to them because of their size. You can find a list of some of my favorites at thefuckitdiet.com/resources. And be careful of the "body positive" media that is anti-fat. Lots of fitness trainers and gurus have adopted the term *body positive*, but will still wax poetic about the importance of losing weight. Make sure your body positivity is also fat positive.

YOU ARE MAKING THE STAKES TOO DAMN HIGH

You are almost definitely making the stakes too high—for everything. Lots of people with a Type A personality, control issues, or perfectionistic tendencies feel that everything is on the brink of falling apart, all the time, unless *we tightly control it.* But unless you hold someone's life *literally* in your hands (read: surgeon), the stakes are not as high as you think. Fitting into your old jeans is not a life-or-death situation.

Actors and writers are always taught to *raise* the stakes to keep the audience interested. The higher the consequences, the more entertaining the scene because, suddenly, everything matters more. You do *not* need to do that in your real life. Let your high-drama stakes stay on your TV screen. Or take an improv or singing class if you're dying to express your dramatic soul. There is no reason to add stress to your real life.

We somehow believe that it is *responsible* to make the stakes high. It feels like we're doing our all and caring deeply. It's how we try and prove that something matters to us. But what we're *actually* doing is running our stress hormones high and hard, making everything miserable, and depleting our energy for years on end. Making the stakes high causes constant low-grade anxiety as well as major spikes in stress.

Think about it, we've been *groomed* to believe that our health and happiness and love life are all *actively on the line with every bite we take.* That somehow this has become a life-or-death situation. Literally, people think they are going to put themselves in an early grave by eating nachos. Fuck that. Now, you *know* that's all hype. It's just fearmongering. That was the goal of those marketing companies hired by drug and diet companies, and now you think you need their stupid programs and their stupid appetite-suppressing lollipops. They made the stakes high, and it makes them a fuck-ton of money—all at your expense.

Looking skinny in your cousin's ugly bridesmaid dress *is not a life-or-death situation*. Honestly—who cares how you look? Who cares about those pictures? Who cares what anyone thinks? FUCK it. Fuck the drama. Fuck the noise. Fuck the diet industry. And now that I think about it, fuck the absurd wedding-industry expectations too.

Hey, I get it. We all just want to be responsible and happy. But we don't need to make the stakes high. Not for your diet, not for impressing fair-weather friends, and not for the size of your pants.

You know what *is* a life-or-death situation? Eating disorders. Not eating *will* kill you. Not eating enough will fuck with your health and your hormones and wreak havoc on your mental health. The stakes *are* high for your mental health and your quality of life, so honestly: fuck it.

HOW TO TRUST WHEN YOU DON'T TRUST

It's easier to lower the stakes if you're willing to have some sort of trust in the bigger picture. And that can be hard, especially if you are used to *not* having any trust. It's really hard for people who believe that it's just them against the world to have trust. Once you are convinced that your past experiences prove that you can't trust, it's hard to convince you otherwise. How can I convince you to trust when you just . . . *don't*? Talk about a limiting belief!

The best way to begin trusting is to start trusting your body. Your body exists to heal you. Your body's signals, cravings, and appetite exist to keep you alive and to take care of you. Your exhaustion, hunger, stress response, and immune response all exist to keep you well. So if you can't have trust in the big picture yet, begin to put some trust in your body.

After years of fighting your body and being at odds with your body, body trust is very hard for lots of people. We are so convinced that our body has let us down. We are convinced that, left to its own devices, our body will betray us time and time again. We think that if we didn't spend most of our energy controlling our appetites with low-calorie roughage and forcing ourselves through depleting workouts, we'd quickly spin out of control.

Every single thing your body has ever done has been to protect you. Our mistake has been believing that there is something wrong with having an appetite or having a body that isn't extremely small.

I can tell you over and over and over that happiness, and health, and worth have nothing to do with your weight. But I still can't make you believe me. I cannot make you trust me. And I cannot make you trust your body.

In order to trust, you have to take a leap of faith. Ask yourself what you really believe to be true in your bones and then act on it. Start listening to your appetite. Start following your cravings. Learn to trust by listening to your body's signals. Your body will not let you down.

You're allowed to be scared, have doubts, and have major limiting beliefs to work through (we all do), but you have to lean into trust. Trust your body, and trust that your existence is more than just one long weight-loss attempt.

LET IT FEEL CRAZY

I sometimes talk about "loving yourself like a psycho." I phrase it that way because some of us think it would be *insane* to like ourselves the way we are right now. *There is no way that any sane person would like themselves the way I am. I don't deserve it. I can't do it. I'd be laughed out of town.*

But why? Why wouldn't you deserve it? One of the most destructive beliefs we learn is that gaining weight makes us ugly, and that being ugly makes us unworthy. So how could we not be afraid of gaining weight? These are fucked-up societal beliefs about weight that are relatively *new*. Less than eighty years ago, they were selling powders, supplements, and all sorts of other things to make women *less* skinny. The kind of body you're supposed to have is always based on whatever social elitism is going on at the time. And unlearning what you've been taught about beauty and worth is more important than anything else you will do on The Fuck It Diet.

Wherever you are is where you have to start. Let go of the blame, the self-hatred, the shame, and the box that you've been put into. Take the time to understand your beliefs about weight, work through them, and let them go.

You are exactly as you should be. You have the right body type.

You do *not* need anyone else to approve of you. You are allowed to live life on your own terms. And you are allowed to feel beautiful even if they have told you that you shouldn't.

And, yeah, it's scary as hell. You have to face all of your biggest fears and shed the identity that used to make you feel safe. But if you aren't willing to do the really scary work, nothing will change, no matter how much research you do on normal eating and body positivity. The desire to heal has to be stronger than your desire to stay in control. The desire to feel discomfort and pain has to become stronger than your desire to numb. The desire to be healthy has to be stronger than your desire to be skinny. It comes down to wanting to be happy more than you want to be beautiful, because if you feel worthy *even if you believe you aren't quite beautiful*, you can't lose.

If you can't love yourself like a psycho just yet, that's okay. For now, just strive to add some compassion into the mix: for yourself, for where you are, and for how hard life is.

I am giving you permission to *let* it feel crazy until it doesn't feel crazy anymore. Accept yourself harder than you think you are allowed to. Radically accept where you are, even if it feels stupid. And eventually you'll realize it isn't stupid at all.

LOVE YOURSELF LIKE A PSYCHO

Write out all the reasons you fear it would be "stupid" to accept yourself. All the things about you that only a stupid person would love. Follow your emotional reactions to your body; this list doesn't have to be rational. Go through the list and see if you can imagine letting yourself be stupid and insane enough to love yourself despite these things.

WAITING TO NOT BE HUNGRY

So many people have the limiting belief that hunger is "a problem that needs to be fixed." This is *such* a common and insidious belief that even people who have been on The Fuck It Diet for a long time realize they are *still* waiting for the day when they won't be hungry anymore. There was a little part of them all along that was waiting to be *healed* from their hunger, as if hunger is a symptom that needs to be cured. Lots of people assume the real goal of all of this is to end up losing our appetites, because it has been so consistently ingrained in us that having an appetite is unhealthy or weak.

It makes sense. Most people start The Fuck It Diet from a "diet mentality," so it's not surprising that so many people automatically hope that The Fuck It Diet will heal us from what we think our problem is: hunger. We think that if we can re-feed ourselves enough, and fix our metabolisms, that eventually we will get to a state where we aren't hungry anymore. And I am here to tell you, again, that is never going to happen.

That is also what lots of us attempted to do with obsessive "intuitive eating." We thought that if we were *really* being intuitive, we would not want to eat. If we were *really* listening, we wouldn't want to eat much at all. Or that somehow magically we'd only ever crave kale.

Now, I do say things like "eat yourself to the other side," and what I mean is: get yourself to the other side of famine mode, where you aren't as ravenous or afraid or *fixated* on food. We don't want to live in a survival famine mode.

However, even once you "get to the other side," where food is just food, you will still have an appetite. Because an appetite is a sign of health and a functioning metabolism. Having an appetite or wanting to eat is not weak, it's being alive. And it is never going to go away. And if it does, go to your doctor, because you might be dying.

IT'S ALLOWED TO BE MESSY

There is no "one and done" method for navigating the way your mind works. Remember, there is no cure-all. And the tools you've learned so far are meant to be used ongoing for the rest of your life. You're a living, breathing human, and you will continually be faced with stressful situations, challenges, and times when you have to figure out where old beliefs are triggering you. You will be finding and releasing limiting beliefs for the rest of your life. The learning is *never* done.

Self-acceptance is not a straight line. You'll have days where you feel awesome and fully healed. And then, boom: self-doubt, self-judgment, and debilitating fear of what other people think. *What am I doing? Am I a crazy person? Why did I think letting go of dieting was a good idea?!?!?! Everyone is judging me and my new pants!!!*

Jumping to the idea that losing weight is the simple answer to your problems is a deep pattern in your brain, as well as our collective cultural consciousness. So when you go through a stressful period, sometimes your brain will sink back into old self-destructive habits. *If I just lose a little weight, everything will fall into place. Maybe then I'd have some goddamn control over the rest of my life.*

Whether we know better or not, impressing people and getting their approval often feels like a really solid way to become safe and happy. Restricting will sometimes still feel like a surefire way to get approval *and* become safe and happy. But in the end, this is just an old coping mechanism that will leave us wanting, needy, hungry, and weird around food.

And if you *do* go back to trying to diet or restrict, you'll find that restriction starts to backfire faster and faster. Eliza reached out to me and said, "I have been on The Fuck It Diet for about ten months now. The level of life intuition and trust in myself that the energy work and food approach has given me has been so transformative. Along the way I have had some slip-ups and returns to restrictive eating, but now the negative effects of dieting are so noticeable, so quickly. I'll immediately feel addicted to food again, and while the lesson is never fun, it helps to show me how important allowing all food is for my intuition, and my physical and mental health."

Don't panic about the panic or feel guilt about the guilt. Sometimes you have to just recommit.

Making mistakes, and taking one step forward and two steps back, is normal. Going back and forth in your confidence doesn't mean this whole thing isn't working, it just means that self-improvement and seeking happiness isn't a straight line. It's messy.

THE THRIVING PART

Your life has been waiting patiently (and boredly) for you to stop worrying about stupid shit. And so it's time to reintroduce you back to your life. Those physical, emotional, and mental phases were the stepping-stones to get you out of survival mode, and into thrive mode. All the work you've done to leave behind your food and weight fixation, to embody your feelings, and to overcome your mental blocks has opened up space so you can start to live your life in a way that is significantly more empowered and intuitive and hopefully: fucking *fun*.

This is where we get to lean into what we really want, what we really think, what we really need, who we really are, and what we are really here for.

We are also going to talk about saying no to people and activities and situations that don't feel good or don't serve us, and start setting some boundaries to make sure we are taking care of and prioritizing ourselves.

They don't teach this stuff in school, and they freaking should.

WHAT DO YOU STAND FOR?

If you zoom way out and think about your life as a blip in the vast story of humanity, what do you think you might be here to do? It *definitely* isn't counting almonds. You are not on this earth to be awesome at low-carb meal prep. You have a purpose way *beyond* stupid, soul-sucking dieting. While that might be obvious, we forget *all of the time* that life just isn't worth all this drama.

At the beginning of my very own Fuck It Diet, I was also having a major existential crisis over my career and "purpose"—which had always been steeped in so much perfectionism, disappointment, and guilt. Months after I decided to heal my relationship to food and weight, I stumbled across the book *The Artist's Way*. And even though the book has nothing to do with eating or body image, it addressed perfectionism and control in a way that seriously changed my life, and informed the way I approached The Fuck It Diet, too. In fact, the brain-dump exercise is adapted from the Morning Pages exercise in *The Artist's Way*.

Perfectionism and *attempted* control are the biggest ways we stifle ourselves. We are so afraid to be imperfect or do a bad job that we would rather just do nothing at all. *The Artist's Way* taught me that anything worth doing is worth doing poorly, because it's not about the end product. The act alone is where the joy is.

This book is in your hands because, years ago, when I was lost and miserable and afraid of tortilla chips, I read a book that encouraged me to make something, even if it was bad. So I made a website, called it The Fuck It Diet, and started writing about what I was learning about the perils of dieting.

Don't let the word *purpose* freak you out. You don't have to do anything epic in order to have a purpose. You don't even have to know what your purpose is, and it can change from month to month or year to year. Purposes can be quiet and unassuming and help anchor you into something a little bit less soul-sucking than whether you look cool enough in your new trendy pants, and they can have a ripple effect just based on how you live your life. Instead of worrying about a bigger purpose, just ask yourself, "What do I stand for, today?"

You don't have to be an extrovert or a fighter to let what you stand for quietly infuse the way you walk through the world. Sure, you can organize marches, create subversive art installations, or be an ambassador to a big charity. But you can also express what you stand for in the way you craft gifts for friends once a year, or make people laugh, or in the flowers you plant. It can be small. It can seem innocuous, but it's not.

When you feel the gravitational pull of the diet hive mind, remember how your life is about so much more than what you look like or how much you weigh. You can look at this from a very spiritual perspective, or from a straightforward, practical perspective. How do you want the way you lived, and the way you connected with people in your life, to affect the world and the next generation? What do you stand for, today?

We have been confused about what is important. We have let our looks and calorie counting eclipse the way we are living our lives. Consider the possibility that finding a better fucking use of energy can help you create a fuller life and help you heal.

WHAT DO YOU STAND FOR?

What do you stand for? When you look back on your life, what are the things you would hope you prioritized? If this is hard to access, imagine an innocent child was affected by the way you think about yourself, and absorbed the things you stand for in this world (and if you are a parent or teacher, this is a reality). How would that clarify what's important to you?

PRACTICAL BOUNDARIES WITH FOOD AND WEIGHT

If your friends and family are known to talk about food and weight (theirs or yours), I recommend you tell them exactly what you're doing. Ask for understanding, support, or at the very least, no comments. And then . . . expect nothing.

Really. *EXPECT NOTHING.* If you expect to be able to convert them to The Fuck It Diet, you will almost certainly be disappointed when one of your senile grandparents asks you loudly in front of everyone if you are thin or fat these days (true story). Or when your uncle tells you he is sure that the Atkins diet works because every year he goes on it again, and every year he loses weight again (another true story).

Again, remember your own journey. You had to come to this on your own. You had to hit your own personal rock bottom with food and weight in order to be able to be open to a radically different alternative. You probably had to experience your own pseudo-intuitive eating before realizing how easy it is to make that into a diet. And you had to go on your own version of the low-carb Atkins diet year after year after year, before you realized that you weren't the broken one.

It can be an uncomfortable conversation, so if you want some guidance on where to start, consider saying something like:

> *Hey, as you probably know, I've been searching for the answer to my food and weight stress for years. It's made me pretty miserable and obsessed, so I am trying something new: I'm learning to eat normally, without obsession or fixation. I'm learning to listen to my body, so I am letting myself eat whatever I want. It's actually working, I am feeling way more normal around food. I've gained some weight, and might gain some more, but it's all part of this process.*
>
> *I am trying to change my relationship to weight too, so I would love and appreciate your support in focusing on talking about other things with me, and not talking about food or weight with me anymore. If you want to know more about the science behind this, I'm happy to share.*

If they sound open and supportive, you can mention *The F*ck It Diet* and the Health at Every Size studies. But truly, don't expect them to jump on board with you. If they do, great. If they don't, great. You still established your boundaries.

If and when they inevitably forget or fail to respect your

request that you focus on things other than weight, you can restate your desires and boundaries:

I know we ***used*** *to talk about weight all the time, but it's really important for my mental and physical health right now that I focus on how I feel, and not on what I weigh. Please don't bring up weight again. I'm working really hard to prioritize my health and happiness, not my weight.*

And if and when they pull the "health card," you can always say:

I appreciate you caring about my health, but I have actually found that the more I focus on weight, the worse my eating and health become. Reading The F*ck It Diet *(or any other book on the same topic) would help explain my new perspective, if you're interested in learning more.*

And if they really, really can't stop bringing up food and weight, you have two choices.

Keep stating what you expect and deserve, and keep your boundaries. *You are not in the wrong.* You have every right to request respect and understanding around your life and health and body choices.

Or stop hanging out with them. Hard to do if they are your mother or coworkers, but boundaries are boundaries are boundaries.

What I recommend you do with people you don't know is ignore them. Remember you were like them once, realize that they are living in an alternate, miserable universe, where there is a weight rat race and we aren't good enough until we prove our worth to ourselves and fit into our preteen jeans. Use rebellion to remind yourself you are your own fucking boss and you are cool and amazing.

The very most important thing to do, in addition to creating practical boundaries, is to figure out what these people are triggering **in you**. If their words keep bringing up stress in you, ask yourself:

Where are you afraid they are right?

What limiting beliefs of YOURS are they forcing you to deal with?

And as always: find those limiting beliefs and release them with Tool #5.

The more confident and assured you become in your own choices and body, the easier it will be to be around people who are not on board with your way of eating and relating to weight. That is the benefit of figuring out where *you fear they may be right* and doing the energy work to help you trust what you are doing and standing in your power.

FRIVOLOUS DOWNTIME

You need, and deserve, frivolous downtime. (Also known as self-care. Also known as mental health time.) Most dieters also have a lot of *workaholic energy*, so in order to fully heal, we have to look beyond just the food and exercise and look at the way you approach . . . everything.

The truth is, your productivity, energy, and innovation will dry up quickly if you aren't taking time to replenish, feed your soul, and frivolously allow time for what you enjoy. *Actually doing things that you enjoy* will replenish you in a way that nothing else really can.

What this means is that becoming aware of the things that you like to do, just for the sake of doing them, is a very *responsible* thing to do for your happiness and productivity at large. Not to bribe you with the paradigm of responsibility and productivity . . . but, hey, if it works, I will.

Of course, the difference is that your happiness is of the utmost importance here. Because if you won't let yourself be happy, what are you really doing? Like, truly, what are you trying to prove? That you can be extremely serious and miserable and productive until the day you die? Bravo. What was the point?

For a long time, I imagined that "self-care" had to be something very feminine and expensive, like getting a facial or taking a long sweaty bubble bath or sniffing aromatherapeutic candles from Anthropologie while wearing beautiful silk pajamas and listening to accordion music.

If that is your ideal version of self-care—DO IT. Be the Amélie you have always dreamed of being. But self-care doesn't have to be quite so *extra*. And if the phrase *self-care* doesn't resonate with you, please substitute another term: mental health days. Mental health time. Mental health afternoons. Remembering how much this is for your mental health, and soul health, makes it clear *how much we need this*.

I define *self-care* as "something you enjoy doing." Something that will replenish you on a deeper level. So ask yourself . . . What do you actually enjoy doing? What have you always been curious about trying? I enjoy the mundaneness of going for walks while listening to music, watching TV and talking about TV story and character arcs with my other TV-loving friends, and slowly but surely learning that if I buy plants that need full sun, and plant them in the shade, they will die (amateur gardening). A few years ago my answer to this question was very different. You're allowed to try things out. You're allowed to change your mind about what you enjoy doing.

Self-care is about being willing to take a time-out and giving yourself what you *actually* need in that moment. It means taking care of yourself in the way that you need most, and prioritizing those needs that you have probably gotten used to ignoring. Sometimes that means napping, sometimes it's canceling plans, sometimes it's making plans with friends or calling someone on the phone, going to therapy, dedicating yourself to the noble art of doing nothing, journaling, sleeping, getting a massage, stretching, taking a walk, watching TV, reading a book, connecting to nature, soaking your feet, listening to music, or eating your favorite food (yes! eating can be self-care!). The options are endless. What *matters* is getting in touch with what you actually need, and being willing to give that to yourself, instead of denying those needs and pushing onward.

The lie-down is self-care, but I wouldn't exactly call it frivolous downtime. Ideally, we have other self-care in *addition* to the lie-down. The lie-down is an adult time-out. This self-care is allowed to engage your brain (or not). This is about finding *other* ways of taking care of ourselves.

The most unhelpful belief we can hold on to is that self-care is selfish or unnecessary. Or that we are weak if we need to take time to ourselves. YOU ARE NO GOOD TO ANYONE, OR TO YOURSELF, IF YOU ARE EXHAUSTED, MISERABLE, AND FEEL TERRIBLE. We all have different ways of recharging, and those needs will change from day to day, week to week, and situation to situation. But no matter who you are, you need to allow time to take care of yourself, and that time doesn't have to look the way you imagine.

I recommend ten to twenty minutes, every day, *minimum*. Again, this is in addition to the lie-down. And take this as far as you want—hey, go ahead and quit your life and take a whole ten years of self-care if you can. But tiny chunks will also do wonders, too. You can ask yourself, "What needs my attention? What might be really nice to do right now?" Go ahead and make a list of five tiny things—is there something on that list you can do for yourself today? Can you do all five?

During this self-care time, you're going to let yourself *slow down*. No pressure to accomplish anything. No goals. No nothin'. This time has *everything* to do with The Fuck It Diet, because even though this whole thing feels like it is all about food, the actual underlying issue is *how hard we are on ourselves*. I am asking you to overhaul the way you do life.

FANTASY SELF-CARE

If there were no obstacles, and you had a magic wand, what kind of self-care would you seek and schedule in your life? Make a list. Let it be a fantasy list. You have billions of dollars and a magic wand.

Once you make the list, see if there is anything on there that you could actually try in your real life, even if you have to adapt it for practicality. For instance, if you would go on a weeklong retreat, how can you bring aspects of the retreat you want to go on into your weekend? Get a massage? Buy some lavender essential oil? Then fucking try it!

IMPROVE YOUR HEALTH WITH NO DIETS OR WEIGHT LOSS OR GYMS

If you are ever freaking out about health, here are just a handful of ways to support your health from a radically antidiet approach. Try these things anytime you need to remind yourself that yes, you do care about yourself, you are taking care of yourself, and everything is going to be okay. And that one day, you will die anyway.

Eat and heal your metabolism.

Let your body gain weight.

Eat carbs.

Sleep a lot.

Do your lie-down.

Say no to things you hate doing.

Say yes to things that sound fun.

Take personal days.

See friends.

Eat probiotics or fermented food.

Take a supplement to support your adrenals and stress hormones.

Breathe deeply.

Stretch.

Find ways to laugh.

Watch or read something life-affirming or heartwarming.

When you have energy, move your body in ways you like.

Go to your weight-neutral doctor.

Get a massage or acupuncture.

Go to a therapist you trust.

Try new foods.

Get out in nature.

Get in the sunshine.

Put plants in your house.

EMOTIONAL AND EXISTENTIAL REST

You know those apps that you forgot were still opened on your computer or phone? The ones that are draining your battery even though you're not using them? That's what the stress and worry over all the things you fear you're doing wrong will do to you: they'll deplete your life force from the background. And you'll keep wondering why you're so exhausted.

A few years into my own Fuck It Diet, I had this radical feeling that I needed to put myself on "two years of rest." I had been letting myself *physically* rest. But it wasn't until a few years in when I realized . . . *Ohhhh. I am still shoulding all over in every OTHER area of my life.* And I realized that all of my *shoulds* were still running the show from the background—and had been for years. They were constantly depleting me, draining my energy and causing anxiety that I couldn't pinpoint. I hadn't actually put my finger on how exhausting limiting beliefs were until then.

The easiest way to find the causes of existential depletion is to ask: *Where am I afraid that I'm not doing life "right"? What are the things that I believe that I should be doing? What are the reasons I believe I am failing, going too slowly, letting people down, letting myself down, or going to die ugly and alone?* All of those things are like a mental (and emotional) marathon, plus they affect our physical bodies with chronic stress. For years, our lives have been fueled by high stress hormones, which deplete our bodies on a very basic physical level.

When I realized this, I knew I needed to give myself existential rest, and I needed it bad. I realized that The Fuck It Diet revolutionized the way I lived, ate, and saw myself. So I decided, *Why not do this with everything?* Why not apply this technique to every area of my life? Because honestly, I was fucking *tired*, and I finally realized that all the limiting beliefs I had about food and my body, I also had about love and career and success and money and the ability to be a constantly fun or productive person. I had all of these beliefs about success and responsibility and how and when I was allowed to relax. And it just made me so . . . tired.

My plan was to do nothing but chill. You know how old-timey doctors would prescribe a relaxing month at the seaside? That's what I needed. But for two years. And I decided on "two years" because it sounded long enough to me to actually be radical. The *point* of my rest was to take all the pressure off myself. To say *fuck it* to the shoulds. I was going to move away from acting, use up the money I had made the year before, and write this book. That's all. But in that first year, I decided to move to a new city, then I searched for and bought a house, moved again into the new house, ran three online programs at once while professionally acting full-time—in rehearsals all day and performing eight shows a week at night—all while I learned the heater in my new house might explode at any minute (thanks for nothing, *inspector*), and had rainwater falling on my head while I slept, all while trying to write this book . . . Basically, life happened. And it was not rest.

So how do you rest when you have no fucking time??? I know that most of you are swamped—you work, are parents, have partners, worry about money, have obligations that you can't shirk, and have your own leaky houses. You can't run away to an island or declare Two Years of Rest. But the good news is, you may as well learn how to rest when life is exhausting, rather than wait until life is easy breezy. Because life is rarely easy breezy for long.

This is how to add in existential rest:

1. **LEARN HOW TO HAVE SOME BOUNDARIES** in your life and learn how to say no to things that you *don't need or want to commit yourself to.*
2. **GET USED TO CARVING OUT LITTLE POCKETS OF TIME** for fun and rest and frivolous downtime, even in a schedule that never seems to let up. You deserve ten minutes here and an hour there, and you need to learn to give that to yourself.
3. **REALIZE THAT EXISTENTIAL REST IS MORE, WELL . . . EXISTENTIAL THAN ANYTHING ELSE.** It's about the way you *look* at your schedule and your obligations and your productivity. It's the way you look at how *deserving* of rest and boundaries you are. It's your way of *looking* at your to-do list and learning to let yourself off the hook. It's taking the pressure off while moving forward. It's your way of understanding how important that vacation is and how important fun and downtime is for your soul and happiness and health. It is the active releasing of shoulds and limiting beliefs and taking the pressure off yourself in any way you can. Rest is partially a state of mind.

It's impossible to tackle everything at once, but in getting to this "thrive" phase, past the distraction of food and body obsession, you now have the space and ability to see all of the other limiting beliefs you are operating under.

I am writing this book now during my two years of rest, because I *want* to, not because my *shoulds* are strangling me. This is a phenomenon similar to taking off pressure to eat a certain way and finding you will be drawn to foods that will be supportive to your body *because* you took the pressure off.

I decided to remove any pressure I used to put on myself to get somewhere. I do not have to get anywhere. I do not have to be anywhere different. I do not have to be "further along." I do not have to be happier or richer or healthier or have things figured out. And neither do you.

WHAT ARE YOU DOING THAT IS DEPLETING YOU?

Who is in your life that is depleting you? What are you allowing that is depleting? What do you wish you said no to more often?

BECOME YOUR OWN DAMN GURU

The goal of this book is to get you healing your relationship with food by trusting your body, feeding your body, *feeling* what it feels like in your body, and letting go of the beliefs that have clouded your own wisdom. You are your best bet in figuring out the best choices for you. Other people's opinions only matter based on how they are resonating with *you* and your own wisdom.

Be wary of anyone who claims they know what is best for your body. Including me. Yes, really! Weigh *everything* against your own wisdom and your own intuition. You are in charge. The tools and writing exercises I've given you in this book are the *simplest* and easiest ways to connect with your own intuition and wisdom. Food will get your body and mind out of survival food fixation. The lie-down tool will give you the chance to slow down. The breathe and feel will let you *notice* and feel what's going on. And the brain dump is an amazing way to get out all the noise and make space for clarity and intuition to pipe up.

Your intuition, unlike your mind, is calm. Your mind is wired for survival, and sure that doom is lurking around every corner. So your mind is essentially just a critical, scared asshole, wailing and complaining and nervous and incessantly berating and filled to the brim with limiting beliefs and shoulds and worries. And then sometimes, through the cracks, peeps a sure, calm nugget of wisdom.

It will take you some time to figure out how to listen to yourself. That's totally expected. You can always try things out and see how things feel. You're allowed to take wrong turns and detours and make mistakes. But still, always allow yourself to be your own authority.

When you are having trouble knowing what is right for you, the answer is almost always *just wait.* Do the brain dump. Lie yourself down in your li'l bed. Maybe take a walk. And wait. You will know what to do. Your intuition is simple, and sure, kind, and calm. It's the calm things you can trust.

WHERE AM I STILL FOLLOWING OTHER PEOPLE'S RULES?

Write out all the things that you do and think and worry about that are based on society's or your family's or your community's beliefs. Notice if any of them are limiting beliefs that need to be released with a separate brain dump. Circle them and add them to your limiting belief master list, to be released later. Then go back through the list and rewrite the rules to suit where you are today. Remember you're in charge. You write the rules for your own life.

THAT'S IT!

Look at you! You made it through the book! Chances are you are still somewhere in the middle of your relationship with food and weight, on your very own Fuck It Diet. But now you have the tools to make some pretty profound shifts away from survival mode, and away from living a life that isn't really yours. Keep using the five tools, not only on your relationship with food and body but in any area of your life that feels muddy and confusing. You can rely on your body's wisdom in *any* area where you crave more internal guidance.

Trust your impulses. Trust your desires and truth. Trust what you know to be true, lower the stakes, and maybe eat some food.

ACKNOWLEDGMENTS

This book was possible because of other people's research, other people's feminism, and other people's defiance that all came before mine. I will be eternally grateful for all the work and research that paved the way for this book.

I am not the first person to write on this topic and I will not be the last. This book was possible because I wrote it during a time when body positivity is becoming more and more mainstream (and during the wave of books with F*ck in the title).

I have to honor:

> All of the scientists and researchers who have gone against the grain and spoken up about food and weight bias, who have published books and open access journals and articles: my work wouldn't be the same without yours.

> All of the fat activists who are writers, athletes, models, comedians, actors, and who have a *way* harder time than me, and who are leading by example: you have taught me

so much. You give the world so much by sharing your experience. I will never stop being thankful to all of you.

All of the therapists, dieticians, nutritionists, nurses, and doctors who are teaching and preaching a food-neutral and weight-neutral approach to health and healing, *you are the ones on the ground,* and your work is essential.

All the people who have taught intuitive eating or non-diet approaches through the years.

All my original students and readers, your trust in TFID and your feedback to me made this book possible. Thank you. Thank you. Thank you.

Elisa, Corey, and Maryellen, thank you for reading the book early; Sam for the photos and dog sitting; Alexis for the magic; Susan and Annie for the nude lipstick; Melanie for the phone calls; Matt for making texting into an art; Margaret and Shane for making me laugh; my parents—who don't like cursing but still support me; Hungry Pigeon for letting me eat breakfast sandwiches while I work; and I would also like to thank my dog, Molly Weasley, for ruining my life and my sheets.

The whole team at Harper Wave who made this book a reality. My editor, Hannah Robinson, and publisher, Karen Rinaldi, thank you for helping me make the best version of this book. And the production team, Brian Perrin, Yelena Nesbit, and Sophia Lauriello, who didn't seem discouraged when we couldn't even put the title of the book in our emails. Thank you, thank you.

Susan Raihofer, my amazing agent, who believed in this book and message even though she had never really been on a diet or struggled with weight. You are truly the best, and you made sure my voice and writing remained un-fucked with.

Emma Lively, who believed in this book and this message and the way I taught: this book would never exist now, in this way, without you. Thank you, thank you for being the *best* creative midwife and book angel. You are one of my favorite people in the world.

NOTES

1. L. Villazon, "Who Would Die First of Starvation—A Fat or a Thin Person?" *Science Focus*, https://www.sciencefocus.com/the-human-body/who-would-die-first-of-starvation-a-fat-or-a-thin-person.
2. M. Nestle, "Why Does the FDA Recommend 2,000 Calories Per Day?," *Atlantic*, August 4, 2011, https://www.theatlantic.com/health/archive/2011/08/why-does-the-fda-recommend-2--000-calories-per-day/243092/.
3. T. Mann, *Secrets from the Eating Lab* (New York: HarperCollins, 2015).
4. L. Bacon and L. Aphramor, *Body Respect* (Dallas: BenBella, 2014).
5. Ibid.
6. "23andMe Releases First-of-its Kind Genetic Weight Report," *23andMe*, March 2, 2017, https://blog.23andme.com/23andme-and-you/23andme-releases-first-of-its-kind-genetic-weight-report/.
7. T. Mann, "You Should Never Diet Again: The Science and Genetics of Weight Loss," *Salon*, April 12, 2015, https://www.salon.com/2015/04/12/you_should_never_diet_again_the_science_and_genetics_of_weight_loss/.
8. Bacon and Aphramor, *Body Respect*.

9. A. Park, "When Exercise Does More Harm than Good," *Time*, February 2, 2015, http://time.com/3692668/when-exercise-does-more-harm-than-good/.
10. R. J. S. Costa, R. M. J. Snipe, C. M. Kitic, and P. R. Gibson, "Systematic Review: Exercise-Induced Gastrointestinal Syndrome—Implications for Health and Intestinal Disease," *Alimentary Pharmacology and Therapeutics* 46 (June 7, 2017), https://doi.org/10.1111/apt.14157.
11. "Too Much Prolonged High-Intensity Exercise Risks Heart Health," news release, American Association for the Advancement of Science, May 14, 2014, https://www.sciencedaily.com/releases/2014/05/140514205756.htm.
12. American Psychological Association, "Work, Stress and Health & Socioeconomic Status," http://www.apa.org/pi/ses/resources/publications/work-stress-health.aspx.
13. M. Seeman and S. Lewis, "Powerlessness, Health and Mortality: A Longitudinal Study of Older Men and Mature Women," *Social Science and Medicine* 41 (August 1995), https://www.ncbi.nlm.nih.gov/pubmed/7481946.
14. V. Felitti et al., "Relationship of Childhood Abuse and Household Dysfunction to Many of the Leading Causes of Death in Adults," *American Journal of Preventive Medicine* 14 (May 1998), https://www.ajpmonline.org/article/S0749-3797(98)00017-8/fulltext.
15. E. Pascoe and L. Richman, "Perceived Discrimination and Health: A Meta-Analytic Review," *Psychological Bulletin* 135 (July 2009), https://www.ncbi.nlm.nih.gov/pmc/articles/PMC2747726/.
16. J. N. Ablin, H. Cohen, M. Eisinger, and D. Buskila, "Holocaust Survivors: The Pain behind the Agony; Increased Prevalence of Fibromyalgia among Holocaust Survivors," *Clinical and Experimental*

Rheumatology 28 (November–December 2010) https://www.ncbi.nlm.nih.gov/pubmed/21176421.

17. K. Schultz, "Are Childhood Trauma and Chronic Illness Connected?," *Healthline*, September 18, 2017, https://www.healthline.com/health/chronic-illness/childhood-trauma-connected-chronic-illness.
18. "Pounding Away at America's Obesity Epidemic," transcript from *Fresh Air*, NPR, May 14, 2012, https://www.npr.org/2012/05/14/152667325/pounding-away-at-americas-obesity-epidemic.
19. F. Q. Nuttall, "Body Mass Index: Obesity, BMI and Health: A Critical Review," *Nutrition Today*, April 7, 2015, https://www.ncbi.nlm.nih.gov/pmc/articles/PMC4890841/.
20. K. Flegal and K. Kalantar-Zadeh, "Overweight, Mortality and Survival," *Obesity* 21 (September 2013), https://onlinelibrary.wiley.com/doi/full/10.1002/oby.20588; M. Harrington, S. Gibson, and R. Cottrell, "A Review and Meta-Analysis of the Effect of Weight Loss on All-Cause Mortality Risk," *Nutrition Research Reviews* 22 (June 2009), https://www.cambridge.org/core/journals/nutrition-research-reviews/article/a-review-and-meta-analysis-of-the-effect-of-weight-loss-on-all-cause-mortality-risk/26226C6DF1BA32BEB00AAC87FC416667.
21. L. Bacon and L. Aphramor, "Weight Science: Evaluating the Evidence for a Paradigm Shift," *Nutrition Journal* 10 (January 2011), https://nutritionj.biomedcentral.com/articles/10.1186/1475-2891-10-9.
22. A. Carroll, *The Bad Food Bible* (New York: Houghton Mifflin Harcourt, 2017).
23. C. Jones, J. Fauber, and K. Fiore, "Slippery Slope: $$ in, Diet Drugs

Out, How Five Drugs Came to Market," *MedPage Today*, April 19, 2015, https://www.medpagetoday.com/special-reports/slipperyslope/51058.

24. P. Marsh and S. Bradley, "Sponsoring the Obesity Crisis," Social Issues Research Centre, June 10, 2004, http://www.sirc.org/articles/sponsoring_obesity.shtml.
25. Amy Erdman Farrell, *Fat Shame: Stigma and the Fat Body in America* (New York: New York University Press, 2011).
26. S. McLeod, "Maslow's Hierarchy of Needs," Simply Psychology, May 21, 2018, https://www.simplypsychology.org/maslow.html.
27. Bacon and Aphramor, *Body Respect*.
28. D. Ciliska, "Set Point: What Your Body Is Trying to Tell You," National Eating Disorder Information Centre, http://nedic.ca/set-point-what-your-body-trying-tell-you.
29. D. Drummond and M. S. Hare, "Dietitians and Eating Disorders," *Canadian Journal of Dietetic Practice and Research* 73 (Summer 2012), special international issue, https://www.ncbi.nlm.nih.gov/pubmed/22668844.
30. M. Weig et al., "Limited Effect of Refined Carbohydrate Dietary Supplementation on Colonization of the Gastrointestinal Tract of Healthy Subjects by *Candida albicans*," *American Journal of Clinical Nutrition* 69 (June 1999), https://www.ncbi.nlm.nih.gov/pubmed/10357735.
31. V. Podgorskiĭ et al., "Yeasts—Biosorbents of Heavy Metals," *Mikrobiolohichnyĭ Zhurnal* 66 (January–February 2004), https://www.ncbi.nlm.nih.gov/pubmed/15104060.
32. N. Barnard, "Does Sugar Cause Diabetes?" *Dr. Barnard's Blog*, August 7, 2017, https://www.pcrm.org/news/blog/does-sugar-cause-diabetes.

33. J. Lott, *In Defense of Sugar* (Venice, FL: Archangel Ink, 2015).
34. J. Hari, *Chasing the Scream* (London: Bloomsbury Circus, 2016).
35. S. Pappas, "Oreos as Addictive as Cocaine? Not So Fast," *LiveScience*, October 16, 2013, https://www.livescience.com/40488-oreos-addictive-cocaine.html.
36. D. Benton, "The Plausibility of Sugar Addiction and Its Role in Obesity and Eating Disorders," *Clinical Nutrition* 29 (June 2010), https://www.ncbi.nlm.nih.gov/pubmed/20056521.
37. M. L. Wolraich, D. Wilson, and J. White, "The Effect of Sugar on Behavior or Cognition in Children: A Meta-Analysis," *Journal of the American Medical Association* 274 (November 22–29, 1995), https://www.ncbi.nlm.nih.gov/pubmed/7474248.
38. Lott, *In Defense of Sugar.*
39. S. Fallon and M. Enig, "Why Butter Is Better," Weston A. Price Foundation, January 1, 2000, https://www.westonaprice.org/health-topics/know-your-fats/why-butter-is-better/.
40. A. Price, "What Is Butyric Acid? 6 Butyric Acid Benefits You Need to Know About," Dr. Axe: Food Is Medicine, June 15, 2017, https://draxe.com/butyric-acid/.
41. M. Satin, "Salt and Our Health," Weston A. Price Foundation, March 26, 2012, https://www.westonaprice.org/health-topics/abcs-of-nutrition/salt-and-our-health/.
42. M. Morris, E. Na, and A. Johnson, "Salt Craving: The Psychobiology of Pathogenic Sodium Intake," *Psychology and Behavior* 94 (August 6, 2018), https://www.ncbi.nlm.nih.gov/pmc/articles/PMC2491403/.
43. J. Stamler, "The INTERSALT Study: Background, Methods, Findings, and Implications." (Feb 1997), https://www.ncbi.nlm.nih.gov/pubmed/9022559.

44. C. Kresser, "Shaking up the Salt Myth: The Human Need for Salt," *Chris Kresser: Let's Take Back Your Health*, April 13, 2012, https://chriskresser.com/shaking-up-the-salt-myth-the-human-need-for-salt/.
45. I. A. Marin et al., "Microbiota Alteration Is Associated with the Development of Stress-Induced Despair Behavior," *Scientific Reports* 7 (March 7, 2017), https://www.nature.com/articles/srep43859.
46. C. Kresser, "How Stress Wreaks Havoc on Your Gut—And What to Do About It," *Chris Kresser: Let's Take Back Your Health*, March 23, 2012, https://chriskresser.com/how-stress-wreaks-havoc-on-your-gut/.
47. C. Gillespie, "Being Overweight Can Actually Be Good for You—Especially After a Heart Attack," *Reader's Digest*, July 23, 2017, https://www.rd.com/health/conditions/can-you-be-overweight-and-healthy/.
48. M. Fabello, "5 Social Theories That Prove Health Is Constructed," *Everyday Feminism*, September 29, 2017, https://everydayfeminism.com/2017/09/proof-that-health-is-constructed/.
49. G. Olwyn, "Part II: What Does BED Really Look Like?," Eating Disorder Institute, July 10, 2015, https://edinstitute.org/paper/2015/7/10/part-ii-what-does-bed-really-look-like.
50. G. Olwyn, "Binges Are Not Binges," Eating Disorder Institute, October 31, 2012, https://edinstitute.org/blog/2012/10/31/bingeing-is-not-bingeing.
51. Bacon and Aphramor, *Body Respect*.
52. Ibid.
53. J. Okwerekwu, "In Treating Obese Patients, Too Often Doctors Can't See Past Weight," *Stat*, June 3, 2016, https://www.statnews.com/2016/06/03/weight-obese-doctors-patients/.

54. S. Cohen et al., "Chronic Stress, Glucocorticoid Receptor Resistance, Inflammation, and Disease Risk," *Proceedings of the National Academy of Sciences* 109 (April 17, 2012), https://doi.org/10.1073/pnas.1118355109.

55. A. Seballo, "Health Benefits of Rest," *Florida Hospital*, February 12, 2014, https://www.floridahospital.com/blog/health-benefits-of-rest.

56. A. Tomiyama et al., "Low Calorie Dieting Increases Cortisol," *Psychosomatic Medicine* 72 (May 2015), https://www.ncbi.nlm.nih.gov/pmc/articles/PMC2895000/.

57. M. Pohl, "Chronic Pain: It's All in Your Head, and It's Real," *Psychology Today*, January 2, 2013, https://www.psychologytoday.com/us/blog/day-without-pain/201301/chronic-pain-it-is-all-in-your-head-and-it-s-real.

58. P. Chödrön, *Comfortable with Uncertainty: 108 Teachings on Cultivating Fearlessness and Compassion* (Boston: Shambhala, 2002).

59. A. Mayyasi, "The Surprising Reason Why Dr. John Harvey Kellogg Invented Corn Flakes," *Priceonomics*, May 17, 2016, https://www.forbes.com/sites/priceonomics/2016/05/17/the-surprising-reason-why-dr-john-harvey-kellogg-invented-corn-flakes/.

60. H. Markel, "The Secret Ingredient in Kellogg's Corn Flakes Is Seventh-Day Adventism," *Smithsonian*, July 28, 2017, https://www.smithsonianmag.com/history/secret-ingredient-kelloggs-corn-flakes-seventh-day-adventism-180964247/.

61. P. Levine, *Waking the Tiger: Healing Trauma* (Berkeley, CA: North Atlantic, 1997).

62. P. Payne, P. Levine, and M. Crane-Godreau, "Somatic Experiencing: Using Interoception and Proprioception as Core Elements of Trauma

Therapy," *Frontiers in Psychology* 6 (February 4, 2015), https://www.ncbi.nlm.nih.gov/pmc/articles/PMC4316402/.

63. C. Pert, *Molecules of Emotion: The Science behind Mind-Body Medicine* (New York: Touchstone, 1997).

64. A. Spiegel, "Mind over Milkshake: How Your Thoughts Fool Your Stomach," *Morning Edition*, NPR, April 14, 2014, https://www.npr.org/sections/health-shots/2014/04/14/299179468/mind-over-milkshake-how-your-thoughts-fool-your-stomach.

65. D. Ingram and M. Mussolino, "Weight Loss from Maximum Body Weight and Mortality: The Third National Heath and Nutrition Examination Survey Linked Mortality File," *International Journal of Obesity* 34 (March 9, 2010), https://www.nature.com/articles/ijo201041.

66. D. Lancer, "Shame: The Core of Addiction and Codependency," *Psych Central*, July 17, 2016, https://psychcentral.com/lib/shame-the-core-of-addiction-and-codependency/.

67. C. Baum, "The Wage Effects of Obesity: A Longitudinal Study," *Health Economics* 13 (September 2004), http://onlinelibrary.wiley.com/doi/10.1002/hec.881/abstract.

68. Bacon and Aphramor, *Body Respect*.

69. J. Ludwig et al., "Neighborhoods, Obesity, and Diabetes—A Randomized Social Experiment," *New England Journal of Medicine* 365 (October 20, 2011), http://www.nejm.org/doi/full/10.1056/NEJMsa1103216.

70. D. Skuse, S. Reilly, and D. Wolke, "Psychosocial Adversity and Growth during Infancy," *European Journal of Clinical Nutrition* 48 (1994): suppl. 1, S113–S130.

ABOUT THE AUTHOR

CAROLINE DOONER is a writer, storyteller, performer, and yoga teacher (who mostly just teaches resting), a former yo-yo dieter, and the creator of The Fuck It Diet. She studied improv at UCB and received a BFA in acting from NYU, and consequently spent her early twenties in leotards, or standing in front of casting directors who told you what kind of person you looked like.

After putting herself through over a decade of obsessive, chronic, miserable dieting and bingeing, her own deep misery and dysfunction led her to investigate, research, and create another way: The Fuck It Diet. Caroline hosts a podcast on her approach to food and teaches online courses in intuitive eating and self-acceptance. This is her first book.

WWW.THEFUCKITDIET.COM